T0223835

Geometric Programming for Design Equation Development and Cost/Profit Optimization

(with illustrative case study problems and solutions)

Third Edition

Synthesis Lectures on Engineering

Geometric Programming for Design Equation Development and Cost/Profit Optimization
(with illustrative case study problems and solutions), Third Edition
Robert C. Creese
2016

Engineering Principles in Everyday Life for Non-Engineers
Saeed Benjamin Niku
2016

A, B, See... in 3D: A Workbook to Improve 3-D Visualization Skills
Dan G. Dimitriu
2015

The Captains of Energy: Systems Dynamics from an Energy Perspective
Vincent C. Prantil and Timothy Decker
2015

Lying by Approximation: The Truth about Finite Element Analysis
Vincent C. Prantil, Christopher Papadopoulos, and Paul D. Gessler
2013

Simplified Models for Assessing Heat and Mass Transfer in Evaporative Towers
Alessandra De Angelis, Onorio Saro, Giulio Lorenzini, Stefano D'Elia, and Marco Medici
2013

The Engineering Design Challenge: A Creative Process
Charles W. Dolan
2013

The Making of Green Engineers: Sustainable Development and the Hybrid Imagination
Andrew Jamison
2013

Crafting Your Research Future: A Guide to Successful Master's and Ph.D. Degrees in Science & Engineering
Charles X. Ling and Qiang Yang
2012

Fundamentals of Engineering Economics and Decision Analysis
David L. Whitman and Ronald E. Terry
2012

A Little Book on Teaching: A Beginner's Guide for Educators of Engineering and Applied Science
Steven F. Barrett
2012

Engineering Thermodynamics and 21st Century Energy Problems: A Textbook Companion for Student Engagement
Donna Riley
2011

MATLAB for Engineering and the Life Sciences
Joseph V. Tranquillo
2011

Systems Engineering: Building Successful Systems
Howard Eisner
2011

Fin Shape Thermal Optimization Using Bejan's Constructal Theory
Giulio Lorenzini, Simone Moretti, and Alessandra Conti
2011

Geometric Programming for Design and Cost Optimization (with illustrative case study problems and solutions), Second Edition
Robert C. Creese
2010

Survive and Thrive: A Guide for Untenured Faculty
Wendy C. Crone
2010

Geometric Programming for Design and Cost Optimization (with Illustrative Case Study Problems and Solutions)
Robert C. Creese
2009

Geometric Programming for Design Equation Development and Cost/Profit Optimization
(with illustrative case study problems and solutions), Third Edition

Robert C. Creese

ISBN: 978-3-031-79375-2   paperback
ISBN: 978-3-031-79376-9   ebook

DOI 10.1007/978-3-031-79376-9

A Publication in the Springer series
SYNTHESIS LECTURES ON ENGINEERING

Lecture #27
Series ISSN
Print 1939-5221   Electronic 1939-523X

Geometric Programming for Design Equation Development and Cost/Profit Optimization

(with illustrative case study problems and solutions)

Third Edition

Robert C. Creese
West Virginia University

SYNTHESIS LECTURES ON ENGINEERING #27

ABSTRACT

Geometric Programming is used for cost minimization, profit maximization, obtaining cost ratios, and the development of generalized design equations for the primal variables. The early pioneers of geometric programming—Zener, Duffin, Peterson, Beightler, Wilde, and Phillips—played important roles in its development. Five new case studies have been added to the third edition. There are five major sections: (1) Introduction, History and Theoretical Fundamentals; (2) Cost Minimization Applications with Zero Degrees of Difficulty; (3) Profit Maximization Applications with Zero Degrees of Difficulty; (4) Applications with Positive Degrees of Difficulty; and (5) Summary, Future Directions, and Geometric Programming Theses & Dissertations Titles. The various solution techniques presented are the constrained derivative approach, condensation of terms approach, dimensional analysis approach, and transformed dual approach. A primary goal of this work is to have readers develop more case studies and new solution techniques to further the application of geometric programming.

KEYWORDS

design optimization, generalized design relationships, cost optimization, profit maximization, cost ratios, constrained derivative, dimensional analysis, condensation of terms, transformed dual, posynominials

Contents

Preface

The purpose of this book is to introduce manufacturing engineers, design engineers, manufacturing technologists, cost engineers, project managers, industrial consultants, business managers, and finance managers to the topic of geometric programming (GP). I was fascinated by the topic when first introduced to it at a National Science Foundation (NSF) Short Course 50 years ago in Austin, Texas in 1967. The topic was only for a day or so of a three week course, but I recognized its potential in the application to riser design in the metal casting industry during the presentation. I was fortunate to have two of the pioneers in GP make the presentations, Doug Wilde of Stanford University and Chuck Beightler of the University of Texas, and had them autograph their book *Foundations of Optimization* for me which I fondly cherish even now.

I finally wrote my first publication using GP on riser design in 1972 and although I had written several journal papers using GP on metal cutting and metal casting riser design problems, I was unable to teach a complete course on the topic. Thus, before I retired, I decided to write a brief book on GP illustrating the basic approach to solving various problems to encourage others to pursue the topic in more depth. After writing the book, I was able to teach a course on GP in 2010 and had each student develop a new problem for a course project. This led to increasing the content of the book as some of the problems were very interesting. Its ability to lead to design and cost relationships in an integrated manner makes this tool essential for engineers, product developers and project managers be more cost competitive in this global market place. It also can be use to maximize profits, which is of interest to finance, accountants, and business leaders.

This book is dedicated to the pioneers of GP such as Clarence Zener, Richard Duffin, Elmor Peterson, Chuck Beightler, Doug Wilde, Don Phillips, and all the other researchers for developing GP. This work is also dedicated to my family members, Natalie and Jennifer; Rob, Denie, Robby and Sammy, and Chal and Joyce.

I also want to recognize those who have assisted me in the reviewing and editing of this work and they are Dr. M. Adithan, Professor Emeritus in Mechanical Engineering at VIT University, Vellore, Tamil Nadu, India and Dr. Deepak Gupta, Associate Professor and Director of Engineering Technology in the Industrial and Manufacturing Engineering Department at Wichita State University. I was encouraged by Dr. Robert M. Stark Emeritus Professor of Mathematics and Emeritus Professor of Civil & Environmental Engineering at the University of Delaware to work on a third edition.

I used the first edition of the book to teach a course on GP at the MS level. The original students, Yi Fang, Mohita Yalamanchi, Srikanth Manukonda, Shri Harsha Chintala, Kartik Ramamoorthy, and Joshua Billups developed various examples from the literature, some of which are included in this newer editions and they taught me a lot about GP. This practice was followed in two subsequent classes and more student developed examples followed. One of the important items discovered during the course was the dimensional analysis technique for the obtaining of additional equations when the degrees of difficulty was positive, and no reference to this technique was found in the literature. Sometimes one learns more when teaching the subject and I have taught the course three times and learned more each time.

I am pleased to have the third edition published in time for the 50th anniversary in 2017 of the publication of *Geometric Programming*, the first book on GP by Clarence Zener, Richard Duffin, and Elmor Peterson. This would also be the 56th anniversary of the first paper on concerning GP in 1961 by Clarence Zener.

Dr. Robert C. Creese, Ph.D., PE, CCC
Professor Emeritus
Industrial & Management Systems Engineering Department
Benjamin W. Statler College of Engineering and Mineral Resources
West Virginia University
December 2016

PART I

Introduction, History, and Theoretical Fundamentals of Geometric Programming

CHAPTER 1

Introduction

1.1 OPTIMIZATION AND GEOMETRIC PROGRAMMING

1.1.1 OPTIMIZATION

Optimization can be defined as the process of determining the best or most effective result utilizing a quantitative measurement system. The measurement unit most commonly used in financial analysis, engineering economics, cost engineering, or cost estimating tends to be currency, such as U.S. Dollars, Euros, Rupees, Yen, Won, Pounds Sterling, Kroner, Kronor, Riyal, Swiss Franc, Shekel, Ruble, Rial, Dinar, or other specific country currency. The optimization may occur in terms of net cash flows, profits, costs, benefit/cost ratio, etc. Other measurement units may be used, such as units of production or production time, and optimization may occur in terms of maximizing production units, minimizing production time, maximizing profits, or minimizing cost. Design optimization determines the best design that meets the desired design constraints at the desired objective, which typically is the minimum cost. Two of the most important criteria for a successful product are to meet all the functional design requirements and to be economically competitive.

There are numerous techniques of optimization methods such as linear programming, dynamic programming, geometric programming, queuing theory, statistical analysis, risk analysis, Monte Carlo simulation, numerous search techniques, etc. Geometric programming is one of the better tools that can be used to achieve the design requirements with a minimal cost objective or a maximum profit objective. The development of the concept of geometric programming started in 1961. Geometric programming can be used not only to provide a specific solution to a problem, but it also can in many instances give a general solution with specific design relationships. These design relationships, based upon the design constants, can then be used for the optimal solution without having to resolve the original problem. A second concept is that the dual solution often gives a constant ratio between the terms of the primal objective function. This permits determination of which terms are most critical and should be focused upon in the design stage. These fascinating characteristics appear to be unique to geometric programming.

This book is at the introductory level to introduce the concept and applications of geometric programming and its capabilities of obtaining generalized design equations. The focus is upon the early approach referred to as posynomial geometric programming (PGP). Many important extensions to geometric programming have been made over the years, such as Algebraic Geometric Programming (AGP), Generalized Geometric Programming (GGP), Fuzzy Geometric Programming (FGP), and Conic Geometric Programming (CGP). Computer software packages have been developed to obtain solutions to specific geometric programming problems, but they do not tend to develop the generalized design equations which are important for designers, developers, and analysts.

1.1.2 GEOMETRIC PROGRAMMING

Geometric programming is a mathematical technique for optimizing positive polynomials, which are called posynomials [1]. This technique has many similarities to the more commonly known linear programming, but has advantages in that:

1. a non-linear objective function can be used;

2. the constraints can be non-linear;

3. the optimal cost value or optimal profit value can be determined with the dual objective function without first determining the specific values of the primal variables;

4. the dual solution often gives the ratio of the various cost terms of the primal objective function to the total cost [2]; and

5. in problems with zero degrees of difficulty, design equations for the primal variables can be obtained in terms of the primal constants. As the degrees of difficulty increase, it is more difficult to obtain the design equations.

Geometric programming can lead to generalized design solutions and specific relationships between a design variable and the constant values. Thus, a cost relationship can be determined in generalized terms when the degrees of difficulty are low, such as zero or one. This major disadvantage is that the mathematical formulation is much more complex than linear programming, and complex problems with high degrees of difficulty are very difficult to solve. It is called geometric programming because it is based upon the arithmetic-geometric inequality where the arithmetic mean is always greater than or equal to the geometric mean. That is:

$$(X_1 + X_2 + \ldots X_n)/n \geq (X_1 * X_2 * \ldots * X_n)^{(1/n)}. \tag{1.1}$$

Another parameter used for evaluation of data is the median, which represents the data point in the middle of a series of numbers which contains an odd number of values or is the average of the two middle values if the series contains an even number of values. The median is much more variable than the arithmetic mean or geometric mean and is based only on one value (middle value) or two values (middle two) and ignores the remaining values.

For example, consider the series of numbers of 6, 12, 20, 22, 50, and 100. Since this series contains an even number of terms, the median is the average of the middle two which is:

$$(20 + 22)/2 = 21.$$

The arithmetic mean would be:

$$(6 + 12 + 20 + 22 + 50 + 100)/6 = 35.$$

The geometric mean would be:

$$(6 * 12 * 20 * 22 * 50 * 100)^{1/6} = 23.26.$$

The arithmetic mean is always greater than the geometric mean. Thus, the methods of determining the "central tendency" of a set of data can give quite different results. The "geometric" in geometric programming indicates that the basis of geometric programming is the geometric mean.

Geometric programming was first presented 66 years ago but has not received the attention similar to that which linear programming has obtained over its history of approximately 80 years. Some of the early historical highlights and achievements of geometric programming developments are presented in the next chapter.

1.2   EVALUATIVE QUESTIONS

1. What is the most common unit of measurement used for optimization?

2. The following series of costs ($) were collected: 2, 4, 6, 8, and 10.

    a. What is the arithmetic mean of the series of costs?

    b. What is the geometric mean of the series of costs?

    b. What is the median of the series of costs?

3. The following series of costs (€) were collected: 20, 50, 100, 500, and 600.

    a. What is the arithmetic mean of the series of costs?

    b. What is the geometric mean of the series of costs?

    c. What is the median of the series of costs?

4. The following series of costs (Rupees) were collected: 5, 7, 8, 12, 16, and 18.

    a. What is the arithmetic mean of the series of costs?

    b. What is the geometric mean of the series of costs?

    c. What is the median of the series of costs?

5. What is the year recognized as the beginning of geometric programming?

1.3 REFERENCES

[1] R. J. Duffin, E. L. Peterson, and C. Zener, *Geometric Programming*, pp. 2–3, John Wiley & Sons, NY, 1967. 4

[2] R. M. Stark and R. L. Nicholls, *Mathematical Foundations for Design*, p. 126, Dover Publications, Inc. Mineola, NY, 2004. 4

CHAPTER 2

Brief History of Geometric Programming

2.1 PIONEERS OF GEOMETRIC PROGRAMMING

Clarence Zener, Director of Science at Westinghouse Electric in Pittsburgh, Pennsylvania, is credited as being the father of geometric programming. In 1961 he published a paper in the *Proceedings of the National Academy of Science* titled "A mathematical aid in optimizing engineering designs" [1], which is considered the first paper on geometric programming. Clarence Zener is better known in electrical engineering for the Zener diode. He later teamed with Richard J. Duffin and Elmor L. Peterson of the Carnegie Institute of Technology (now Carnegie-Mellon University) to write the first book on geometric programming, *Geometric Programming* in 1967 [2]. A report by Professor Douglas Wilde and graduate student Ury Passey on "Generalized Polynomial Optimization" was published in August 1966. Professor Douglas Wilde of Stanford University and Professor Charles Beightler of the University of Texas included a chapter on geometric programming in their text *Foundations of Optimization* [3]. I attended an Optimization Short Course at the University of Texas in August 1967 and that is when I first discovered and became interested in geometric programming. I was fascinated by the property that the dual variables gave the cost ratios of the primal objective function and was independent of the values of the constants in the primal objective function. I realized at that time that geometric programming could be used for the metal casting riser design problem, and I published a paper on this in 1971 [4].

Other early books by these leaders were *Engineering Design by Geometric Programming* by Clarence Zener in 1971 [5], *Applied Geometric Programming* by C. S. Beightler and D. T. Phillips in 1976 [6], and the second edition of *Foundations of Optimization* by C. S. Beightler, D. T. Phillips, and D. Wilde in 1979 [7]. Many of the initial applications were in the area of transformer design, as Clarence Zener worked for Westinghouse Electric, and in the area of chemical engineering, which was the area emphasized by Beightler and Wilde. Several graduate students played an important role in the development of geometric programming, namely Elmor Peterson at the Carnegie Institute of Technology and Ury Passy and Mordecai Avriel at Stanford University.

Geometric programming has attracted a fair amount of interest, and a list of the various thesis and dissertations published that have either mentioned or focused on geometric programming is presented in Chapter 25. Websites on geometric programming [8] have appeared with additional interesting applications.

2.2   EVALUATIVE QUESTIONS

1. Who is recognized as the father of geometric programming?

2. When was the first book published on geometric programming and what was the title of the book?

3. Which three universities played an important role in the development of geometric programming?

2.3   REFERENCES

[1] C. Zener, A Mathematical aid in optimizing engineering design, *Proc. of the National Academy of Science*, Vol. 47, p. 537, 1961. 7

[2] R. J. Duffin, E. L. Peterson and C. Zener, *Geometric Programming*, John Wiley & Sons, NY, 1967. 7

[3] D. J. Wilde and C. S. Beightler, *Foundations of Optimization*, Prentice-Hall, Englewood Cliffs, NJ, 1967. 7

[4] R. C. Creese, Optimal riser design by geometric programming, *AFS Cast Metals Research Journal*, Vol. 7, pp. 118–121, 1971. 7

[5] C. Zener, *Engineering Design by Geometric Programming*, John Wiley & Sons, NY, 1971. 7

[6] C. S. Beightler and D. T. Phillips, *Applied Geometric Programming*, John Wiley & Sons, NY, 1976. 7

[7] C. S. Beightler, D. T. Phillips, and D. J. Wilde, *Foundations of Optimization*, 2nd ed., Prentice Hall, Englewood Cliffs, NJ, 1979. 7

[8] http://www.mpri.lsu.edu/textbook/Chapter3.htm (Chapter 3, Geometric Programming) (site visited 5-20-09). 7

Theoretical Fundamentals

3.1 PRIMAL AND DUAL FORMULATIONS

Geometric programming requires that the expressions used are posynomials, and it is necessary to distinguish between functions, monomials, and posynomials [1]. Posynomial is meant to indicate a combination of "positive" and "polynomial" and implies a "positive polynomial." Examples of functions, which are monomials, are:

$$5x, \quad 0.25, \quad 4x^2, \quad 2x^{1.5}y^{-0.15}, \quad 160, \quad 65x^{-15}t^{10}z^2.$$

Examples of posynomials are, which are monomials or sums of monomials, are:

$$5 + xy, \quad (x + 2YZ)^2, \quad x + 2y + 3z + t, \quad x/y + z35x^{1.5} + 72Y^3.$$

Examples of expressions which are not posynomials are:

-1.5 (negative sign)

$(2 + 2yz)^{3.2}$ (fractional power of multiple term which cannot be expanded)

$x - 2y + 3z$ (negative sign)

$x + \sin(x)$ (sine expression can be negative).

The coefficients of the constants must be positive, but the coefficients of the exponents can be negative. The use of signum functions permitted the use of negative coefficients on the constant terms.

The mathematics of geometric programming are rather complex, however the basic equations are presented and followed by an illustrative example. The theory of geometric programming is presented in more detail in some of the references [1–5] listed at the end of the chapter. The primal problem is complex, but the dual version is much simpler to solve. The dual is the version typically solved, but the relationships between the primal and dual are needed to determine the specific values of the primal variables. The primal problem is formulated as:

$$Y_m(X) = \sum_{T=1}^{T_m} \sigma_{mt} C_{mt} \prod_{n=1}^{N} X_n^{a_{mtn}}; \quad m = 0, 1, 2, \ldots, M. \tag{3.1}$$

With $\sigma_{mt} = \pm 1$ and $C_{mt} > 0$

and $Y_m(X) \le \sigma_m$ for $m = 1, \ldots, M$ for the constraints

where $C_{mt} = $ positive constant coefficients in the cost and constraint equations

and $Y_m(X) = $ primal objective function

and $\sigma_{mt} = $ signum function used to indicate sign

of term in the equation (either $+1$ or -1).

The dual is the problem formulation that is typically solved to determine the dual variables and value of the objective function. The dual objective function is expressed as:

$$d(\omega) = \sigma \left[\prod_{m=0}^{M} \prod_{t=1}^{T_m} (C_{mt}\omega_{m0}/\omega_{mt})^{\sigma_{mt}\omega_{mt}} \right]^{\sigma} \qquad m = 0, 1, \ldots, M \text{ and } t = 1, 2, \ldots, T_m \quad (3.2)$$

where

$$\sigma = \text{ signum function } (\pm 1)$$
$$C_{mt} = \text{ constant coefficient}$$
$$\omega_{m0} = \text{ dual variables from the linear inequality constraints}$$
$$\omega_{mt} = \text{ dual variables of dual constraints, and}$$
$$\sigma_{mt} = \text{ signum function for dual constraints}$$

and by definition

$$\omega_{00} = 1. \qquad (3.3)$$

The dual is formulated from four conditions:

(1) a normality condition

$$\sum_{T=1}^{T_m} \sigma_{0t}\omega_{0t} = \sigma \qquad \text{where } \sigma = \pm 1 \qquad (3.4)$$

where

$$\sigma_{0t} = \text{ signum of objective function terms}$$
$$\omega_{0t} = \text{ dual variables for objective function terms.}$$

(2) N orthogonal conditions

$$\sum_{m=0}^{M} \sum_{t=1}^{T} \sigma_{mt}a_{mtn}\omega_{mt} = 0 \qquad (3.5)$$

where

$$\sigma_{mt} = \text{ signum of constraint term}$$
$$a_{mtn} = \text{ exponent of design variable term}$$
$$\omega_{mt} = \text{ dual variable of dual constraint.}$$

(3) T non-negativity conditions (dual variables must be positive)

$$\omega_{mt} \geq 0 \qquad m = 0, 1, \ldots, M \text{ and } t = 1, 2, \ldots, T_m. \tag{3.6}$$

(4) M linear inequality constraints

$$\omega_{mo} = \sigma_m \sum_{t=1}^{T_m} \sigma_{mt} \omega_{mt} \geq 0. \tag{3.7}$$

The dual variables, ω_{mt}, are restricted to being positive, which is similar to the linear programming concept of all variables being positive. If the number of independent equations and variables in the dual are equal, the degrees of difficulty are zero. The degrees of difficulty are the difference between the number of dual variables and the number of independent linear equations; and the greater the degrees of difficulty, the more difficult the solution. The degrees of (D) can be expressed as:

$$D = T - (N + 1), \tag{3.8}$$

where

$\qquad T = $ total number of terms (of primal)

$\qquad N = $ number of orthogonality conditions plus normality condition

$\qquad\qquad$ (which is equivalent to the number of primal variables).

Once the dual variables are found, the primal variables can be determined from the relationships:

$$C_{ot} \prod_{n=1}^{N} X_n^{a_{mtn}} = \omega_{ot} \sigma d(\omega) \qquad t = 1, \ldots, T_o, \tag{3.9}$$

and

$$C_{mt} \prod_{n=1}^{N} X_n^{a_{mtn}} = \omega_{mt}/\omega_{mo} \qquad t = 1, \ldots, T_o \text{ and } m = 1, \ldots, M. \tag{3.10}$$

The theory may appear to be overwhelming with all the various terms, but various examples will be presented in the following chapters to illustrate the application of the various equations.

There are three sections of examples, the first section considering cost minimization examples with zero degrees of difficulty, the second section considering profit maximization examples with zero degrees of difficulty, and then the third section considering cost and profit examples with one or more degrees of difficulty and presenting various approaches to solving the problem. Problems with zero degrees of difficulty typically will have a linear dual formulation with an equal number of equations and dual variables and can be solved relatively easily. When there are zero degrees of difficulty, a global solution is obtained rather than a local solution [1, 3]. If the degrees of difficulty are negative, the problem has more variables than terms and the problem is not considered and needs to be modified.

When there are more than one degree of difficulty, the solution is much more difficult. Some of the approaches to solve these problems are:

1. Finding additional equations so the number of variables and number of equations are equal, but the additional equations are usually non-linear. The additional equations can be determined by either: (a) dimensional analysis of the primal dual relationships or (b) by substitution techniques.

2. Express the dual in terms of only one dual variable by substitution and take the derivative of the logarithmic form of the dual, set it to zero, and obtain the value of the unknown dual variable.

3. Condensation techniques by combining two or more of the primal terms to reduce the number of dual variables. This technique gives an approximate optimal solution and will be illustrated.

4. Transformed Problem Approach. This technique is often used on maximization problems to transform signomial problems into posynomial problems. It is also used to have the terms of the transformed objective function as a function of a single variable to make the solution easier.

3.2 EVALUATIVE QUESTIONS

1. What version of the geometric problem formulation is solved for the objective function and why?

2. What values can the signum function have?

3. How are the primal variables determined?

4. Which of the following terms are posynomials?

   a) 3.4, b) $4x$ c) $5xy$ d) $4x^{2.1}$ e) $(x + 2)^4$
   f) $5x - 3$ g) $6x^{-2.4}$ h) e^{-3} i) $3x + 4y + 5z^{-1.4}$ j) $(x + 4)^{2.2}$
   k) $(x - y + 3)^2$ l) $\cot y$ m) $x/2y$ n) $5l^{-2}t^5$.

3.3 REFERENCES

[1] D. J. Wilde and C. S. Beighler, *Foundations of Optimization*, Prentice-Hall, Englewood Cliffs, NJ, 1967. 9, 11

[2] S. Boyd, S.-J. Kim, L. Vandengerghe, and A. Hassibi, A tutorial on geometric programming, *Optimization and Engineering*, 8(1), pp. 67–127, Springer, Germany, 2007.

[3] R. J. Duffin, E. L. Peterson, and C. Zener, *Geometric Programming*, John Wiley & Sons, NY, 1967. 11

[4] C. Zener, *Engineering Design by Geometric Programming*, John Wiley & Sons, NY, 1971.

[5] C. S. Beightler, D. T. Phillips, and D. J. Wilde, *Foundations of Optimization*, 2nd ed., Prentice Hall, Englewood Cliffs, NJ, 1979. 9

PART II

Geometric Programming Cost Minimization Applications with Zero Degrees of Difficulty

CHAPTER 4

The Optimal Box Design Case Study

4.1 INTRODUCTION

The optimal box design problem is a relatively easy problem which illustrates the procedure for solving a geometric programming problem with zero degrees of difficulty. This was a problem initially considered in a goal programming class and since it was not linear it was difficult to solve. However, it is a very easy problem to solve by geometric programming and helped get students interested in the topic of geometric programming. It also indicates the importance of a general solution which is possible with geometric programming; that is, design formulas for the box dimensions can be developed which will give the answers without needing to resolve the problem if the costs or box volume changes.

4.2 THE OPTIMAL BOX DESIGN PROBLEM

A box manufacturer wants to determine the optimal dimensions for making boxes to sell to customers. The cost for production of the sides is C_1 (\$ 2/sq ft) and the cost for producing the top and bottom is C_2 (\$ 3/sq ft) as more cardboard is used for the top and bottom of the boxes. The volume of the box is to be set at a limit of "V" (4 ft^3) which can be varied for different customer specifications. If the dimensions of the box are W for the width, H for the box height, and L for the box length, what should the dimensions be based upon the cost values and box volume?

The problem is to minimize the box cost for a specific box volume. The primal objective function is:

$$\text{Minimize} \quad \text{Cost}(Y) = 2C_2WL + 2C_1H(W + L), \tag{4.1}$$

$$\text{Subject to} \quad WLH \geq V. \tag{4.2}$$

However, in geometric programming the inequalities must be written in the form of $\leq$, and the right-hand side must be ± 1. Thus, the primal constraint becomes:

$$\text{Minimize} \quad \text{Cost}(Y) = 2C_1HW + 2C_1HL + 2C_2WL, \tag{4.3}$$

$$\text{Subject to} \quad -WHL/V \leq -1. \tag{4.4}$$

From the coefficients and signs, the signum values for the dual are:

$$\begin{aligned}
\sigma_{01} &= 1 \\
\sigma_{02} &= 1 \\
\sigma_{03} &= 1 \\
\sigma_{11} &= -1 \\
\sigma_{1} &= -1.
\end{aligned}$$

Thus, the dual formulation from Chapter 3 would be:

$$\text{Objective Function} \quad \omega_{01} + \omega_{02} + \omega_{03} \qquad\qquad = 1 \tag{4.5}$$

$$L \text{ terms} \qquad\qquad \omega_{02} + \omega_{03} - \omega_{11} = 0 \tag{4.6}$$

$$H \text{ terms} \qquad \omega_{01} + \omega_{02} \qquad - \omega_{11} = 0 \tag{4.7}$$

$$W \text{ terms} \qquad \omega_{01} \qquad + \omega_{03} - \omega_{11} = 0. \tag{4.8}$$

The degrees of difficulty (D) are equal to:

$$D = T - (N + 1) = 4 - (3 + 1) = 0. \tag{4.9}$$

Where

T = total number of terms of primal and

N = number of orthogonality conditions plus normality condition or the number of primal variables (H, W, and L are the 3 primal variables).

Thus, one has the same number of variables as equations, so this can be solved by simultaneous equations as these are linear equations.

Using Equations (4.5)–(4.8), the values for the dual variables are found to be:

$$\begin{aligned}
\omega_{01} &= 1/3 \\
\omega_{02} &= 1/3 \\
\omega_{03} &= 1/3 \\
\omega_{11} &= 2/3
\end{aligned}$$

and by definition

$$\omega_{00} = 1.$$

The three dual variables of the dual objective function are all 1/3 which implies that the three terms of the primal objective will contribute equally to the optimal value of the primal objective function. The two width sides, the two length sides, and the top/bottom will each contribute one-third to the total cost.

Using the linearity inequality equation:

$$\omega_{10} = \omega_{mt} = \sigma_m \sum \sigma_{mt}\omega_{mt}$$

$$= (-1) * (-1 * 2/3) = 2/3 > 0 \qquad \text{where } m = 1 \text{ and } t = 1. \tag{4.10}$$

The dual objective function can be found from:

$$d(\omega) = \sigma \left[\prod_{m=0}^{M} \prod_{t=1}^{T_m} (C_{mt}\omega_{mo}/\omega_{mt})^{\sigma_{mt}\omega_{mt}} \right]^{\sigma} \tag{4.11}$$

$$d(\omega) = 1 \left[\{(2C_1 * 1)/(1/3)\}^{(1)*(1/3)} * \{(2C_1 * 1)/(1/3)\}^{(1)*(1/3)} * \right.$$

$$\left. \{(2C_2 * 1)/(1/3)\}^{(1)*(1/3)} * \{((1/V) * (2/3))/(2/3)\}^{(-1)*(2/3)} \right]^1$$

$$= 1 \left[\{(6C_1)^{1/3}\} * \{(6C_1)^{1/3}\} * \{(6C_2)^{1/3}\} * \{(1/V)^{-2/3}\} \right]$$

$$= 6C_1^{2/3}C_2^{1/3}V^{2/3} \tag{4.12}$$

$$= 6 \left(2^{2/3}3^{1/3}4^{2/3} \right) = \$34.62.$$

The solution has been determined without finding the values for L, W, or H. Also note that the dual expression is expressed in constants and thus the answer can be found without having to resolve the entire problem as one only needs to use the new constant values. To find the values of L, W, and H, one must use the primal-dual equations which are

$$C_{ot} \prod_{n=1}^{N} X_n^{a_{mtn}} = \omega_{ot}\sigma d(\omega) \qquad t = 1, \ldots T_o, \tag{4.13}$$

and

$$C_{mt} \prod_{n=1}^{N} X_n^{a_{mtn}} = \omega_{mt}/\omega_{mo} \qquad t = 1, \ldots T_o \quad \text{and} \quad m = 1, \ldots, M. \tag{4.14}$$

Using Equation (4.13) the relationships are

$$2C_1HW = \omega_{01}d(\omega) = d(\omega)/3$$
$$2C_1HL = \omega_{02}d(\omega) = d(\omega)/3$$
$$2C_2WL = \omega_{03}d(\omega) = d(\omega)/3.$$

Combining the first two of these relationships one obtains

$$W = L. \tag{4.15}$$

Combining the last two to the previous line of these relationships one obtains

$$H = (C_2/C_1)W. \tag{4.16}$$

Since $V = HWL = (C_2/C_1)LLL = (C_2/C_1)L^3$. Thus,

$$L = [V\,(C_1/C_2)]^{1/3} \tag{4.17}$$

$$W = L = [V\,(C_1/C_2)]^{1/3} \tag{4.18}$$

$$H = (C_2/C_1)L = (C_2/C_1)[V(C_1/C_2)]^{1/3} = \left[V\,(C_2^2/C_1^2)\right]^{1/3}. \tag{4.19}$$

Since W and L are equal, this suggests the optimal box shape is a square box when volume is the only constraint. The specific values for this particular problem would be:

$$L = [4(2/3)]^{1/3} = 1.387\ \text{ft}$$
$$W = L = 1.387\ \text{ft}$$
$$H = \left[4\,(3^2/2^2)\right]^{1/3} = 2.080\ \text{ft}.$$

The volume of the box, $LWH = (1.387)(1.387)(2.080) = 4.0\ \text{ft}^3$, which was the minimum volume required for the box. Note that the height is greater than the width or length as the cost for the top/bottom is greater than the cost of the sides. To verify the results, the parameters are used in the primal objective function Equation (4.3) to make certain the solution obtained is the same.

$$\text{Cost}\,(Y) = 2C_1 HW + 2C_1 HL + 2C_2 WL \tag{4.3}$$

$$Y_0 = 2C_1 \left[V\,(C_2^2/C_1^2)\right]^{1/3} [V(C_1/C_2)]^{1/3}$$
$$+ 2C_1 \left[V\,(C_2^2/C_1^2)\right]^{1/3} [V(C_1/C_2)]^{1/3}$$
$$+ 2C_2 [V(C_1/C_2)]^{1/3} [V(C_1/C_2)]^{1/3}$$
$$Y_0 = 2C_1^{2/3}\left(V^{2/3}\right)C_2^{1/3} + 2C_1^{2/3}\left(V^{2/3}\right)C_2^{1/3} + 2C_1^{2/3}\left(V^{2/3}\right)C_2^{1/3}$$
$$Y_0 = 6C_1^{2/3}C_2^{1/3}V^{2/3}. \tag{4.20}$$

The design equations for the box dimensions are L, W, and H are Equations (4.17)–(4.19) and the box design cost equation is Equation (4.20).

The expressions for Equation (4.12) from the primal and Equation (4.20) from the dual are equivalent. The geometric programming solution is in general terms, and thus can be used for any values of C_1, C_2, and V. This ability to obtain general relationships makes the use of geometric programming a very valuable tool for cost engineers.

If we use the values of $C_1 = 2\$/\text{ft}^2, C_2 = 3\$/\text{ft}^2$ and $V = 4\ \text{ft}^3$ in Equation (4.3) one obtains:

$$\text{Cost}(Y) = 2C_1 HW + 2C_1 HL + 2C_2 WL$$
$$= 2 \times 2 \times 2.080 \times 1.387 + 2 \times 2 \times 2.080 \times 1.387 + 2 \times 3 \times 1.387 \times 1.387$$
$$= 11.54 + 11.54 + 11.54$$
$$= 34.62.$$

Note that the contribution of the three terms of the primal are equivalent, which is what the dual variables indicated by the 1/3 values.

Now let us assume that the cost values change so that $C_1 = 6$ and $C_2 = 1$, that is C_1 has increased greatly and C_2 is greatly reduced. The dual values will contribute the same ratio to the objective function, but the values will be different. Using Equations (4.17), (4.18), and (4.19) the values of L, W, and H and Equation (4.3) for the primal objective function one obtains:

$$L = [4(6/1)]^{1/3} = 2.884 \text{ ft}$$

$$W = L = 2.884 \text{ ft}$$

$$H = \left[4\left(1^2/6^2\right)\right]^{1/3} = 0.4807 \text{ ft}$$

$$V = 2.884 \times 2.884 \times 0.4807 = 4.0$$

$$\text{Cost}(Y) = 2C_1HW + 2C_1HL + 2C_2WL$$

$$= 2 \times 6 \times 0.4807 \times 2.884 + 2 \times 6 \times 0.4807 \times 2.884 + 2 \times 1 \times 2.884 \times 2.884$$

$$= 16.64 + 16.64 + 16.64$$

$$= 49.92.$$

Note that although the total cost changed with the changed cost components, each of the three terms contributed the same amount to the objective function. The shape of the box changed considerably, but the volume is still the same. If the dual variables for the dual objective function were considerably different, the design focus would be on the term with the highest dual variable as it would have the highest impact on the total cost.

4.3   EVALUATIVE QUESTIONS

1. A large box is to be made with the values of $C_1 = 4$ Euros/m^2, $C_2 = 4$ Euros/m^2, and $V = 8$ m^3. What is the cost (Euros) and the values of H, W, and L?

2. The cost of the top and bottom (C_2) is increased to 6 Euros/m^2 and what is the increase in the box cost and the change in box dimensions?

3. The box is to be open (that is there is no top). Determine the expressions for H, W, and L for an open box and determine the cost if $C_1 = 4$ Euros/m^2, $C_2 = 6$ Euros/m^2, and $V = 8$ m^3. Compare the results with Problem 2 and discuss the differences in the formulas, dimensions, and costs. Is there a change as to what part of the box contributes most or least to the design?

CHAPTER 5

Trash Can Case Study

5.1 INTRODUCTION

Various case studies are used to illustrate the different applications of geometric programming as well as to illustrate the different conditions that must be evaluated in solving the problems. The second case study, the trash can case study, is easy to solve and has zero degrees of difficulty. It is similar to the box problem, but involves a different shape. This problem was developed by one of the students in the course. The solution is provided in detail, giving the general solution for the problem in addition to the specific solution. These examples are provided so that the readers can develop solutions to specific problems that they may have and to illustrate the importance of the generalized solution.

5.2 THE OPTIMAL TRASH CAN DESIGN PROBLEM

Bjorn of Sweden has entered into the trash can manufacturing business and he is making cylindrical trash cans and wants to minimize the material cost. The trash can is an open cylinder and designed to have a specific volume. The objective will be to minimize the total material cost of the can. Figure 5.1 is a sketch of the trash can illustrating the design parameters used, the radius and the height of the trash can. The bottom and sides can be of different costs as the bottom is typically made of a thicker material. The primal objective function is:

$$\text{Minimize} \quad \text{Cost}(Y) = C_1 \pi r^2 + C_2 2\pi rh \qquad (5.1)$$
$$\text{Subject to} \quad V = \pi r^2 h, \qquad (5.2)$$

where:

r = radius of trash can bottom

h = height of trash can

V = volume of trash can

C_1 = material constant cost of bottom material of trash can

C_2 = material constant cost of side material of trash can.

The constraint must be written in the form of an inequality, so

$$\pi r^2 h \geq V. \qquad (5.3)$$

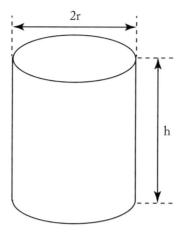

Figure 5.1: Trash can parameters r and h.

And it must be written in the $\leq$ form, so it becomes

$$- \pi r^2 h/V \leq -1. \tag{5.4}$$

Thus, the primal problem is:

$$\text{Minimize} \quad \text{Cost}(Y) = C_1 \pi r^2 + C_2 2 \pi r h \tag{5.5}$$

$$\text{Subject to} \quad - \pi r^2 h/V \leq -1. \tag{5.6}$$

From the coefficients and signs, the signum values for the dual are:

$$\sigma_{01} = \quad 1$$
$$\sigma_{02} = \quad 1$$
$$\sigma_{11} = -1$$
$$\sigma_1 \ = -1 \ .$$

The dual formulation is:

$$\text{Objective Function} \quad \omega_{01} + \omega_{02} \qquad \quad = 1 \tag{5.7}$$

$$r \text{ terms} \qquad 2\omega_{01} + \omega_{02} - 2\omega_{11} = 0 \tag{5.8}$$

$$h \text{ terms} \qquad \qquad \omega_{02} - \ \omega_{11} = 0 \ . \tag{5.9}$$

The degrees of difficulty are equal to:

$$D = T - (N + 1) = 3 - (2 + 1) = 0. \tag{5.10}$$

Using theses equations, the values of the dual variables are found to be:

$$\omega_{01} = 1/3$$
$$\omega_{02} = 2/3$$
$$\omega_{11} = 2/3$$

and by definition

$$\omega_{00} = 1.$$

Using the linearity inequality equation,

$$\omega_{10} = \omega_{mt} = \sigma_m \sum \sigma_{mt}\omega_{mt} = (-1) * (-1 * 2/3) = 2/3 > 0 \quad \text{where } m = 1 \text{ and } t = 1.$$

The value of ω_{02} is twice that of ω_{01} so the second term of the primal objective function will be twice that of the first term. That implies the trash can side will be twice the cost of the bottom of the trash can. This ratio is independent of the specific material cost values of C_1 and C_2.

The objective function can be found using the dual expression:

$$Y = d(\omega) = \sigma \left[\prod_{m=0}^{M} \prod_{t=1}^{T_m} (C_{mt}\omega_{mo}/\omega_{mt})^{\sigma_{mt}\omega_{mt}} \right]^{\sigma} \tag{5.11}$$

$$= 1 \left[\left[\{(\pi C_1 * 1/(1/3))\}^{(1*1/3)} \right] \left[\{\pi C_2 * 1/(2/3))\}^{(1*2/3)} \right] \right.$$

$$\left. \left[\{(\pi/V) * ((2/3)/(2/3)\}^{(-1*2/3)} \right] \right]^{1}$$

$$= 3\pi^{1/3} C_1^{1/3} C_2^{2/3} V^{2/3}. \tag{5.12}$$

The values for the primal variables can be determined from the relationships between the primal and dual as:

$$C_1 \pi r^2 = \omega_{01} Y = 1/3 * Y \tag{5.13}$$

and

$$C_2 2\pi rh = \omega_{02} Y = 2/3 * Y. \tag{5.14}$$

Dividing these expressions and reducing terms one can obtain:

$$r = (C_2/C_1) * h. \tag{5.15}$$

Setting

$$V = \pi r^2 h. \tag{5.16}$$

And using the last two equations one can obtain

$$h = \left((V/\pi)\left(C_1^2/C_2^2\right)\right)^{1/3} \tag{5.17}$$

and

$$r = \left((V/\pi)\left(C_2/C_1\right)\right)^{1/3}. \tag{5.18}$$

Equations (5.17) and (5.18) are considered as the design equations for the trash can. Using Equations (5.17) and (5.18) in Equation (5.5) for the primal, one obtains:

$$Y = C_1 \pi r^2 + C_2 2\pi r h \tag{5.5}$$
$$= 3\pi^{1/3} C_1^{1/3} C_2^{2/3} V^{2/3}. \tag{5.19}$$

Note that Equations (5.19) and (5.12) are identical, which is what should happen, as the primal and dual objective functions must be identical. Equation (5.19) is considered to be the design cost equation for the trash can.

An important aspect of the dual variables is that they indicate the effect of the terms upon the solution. The values of $\omega_{02} = 2/3$ and $\omega_{01} = 1/3$ indicate that the second term has twice the cost impact as the first term in the primal. For example, if $C_1 = 9$ \$/sq ft, $C_2 = 16$ \$/sq ft, and $V = 4\pi = 12.57$ cubic feet, then

$$h = \left((V/\pi)\left(C_1^2/C_2^2\right)\right)^{1/3} = \left((4\pi/\pi)\left(9^2/16^2\right)\right)^{1/3} = (4*81/256)^{1/3} = 1.082 \text{ ft}$$

and

$$r = \left((V/\pi)(C_2/C_1)\right)^{1/3} = \left((4\pi/\pi)(16/9)\right)^{1/3} = (4*16/9)^{1/3} = 1.923 \text{ ft}.$$

Note that

$$V = \pi r^2 h = 3.1416 * 1.082 \text{ ft} * (1.923 \text{ ft})^2 = 12.6 \text{ ft}^3$$

and

$$Y = C_1 \pi r^2 + C_2 2\pi r h = 9 * 3.14 * 1.923^2 + 16 * 2 * 3.14 * 1.923 * 1.082$$
$$= \$104.5 + \$209.0$$
$$= \$313.5.$$

The contribution of the second term is twice that of the first term, which is what is predicted by the value of the dual variables. This occurs regardless of the values of the material cost constants used, and this is an important concept for cost analysis. Now consider reducing the cost of the second term from 16 \$/sq ft to 8 \$/sq ft and what is the effect on the total cost and the dimensions.

$$h = \left((V/\pi)\left(C_1^2/C_2^2\right)\right)^{1/3} = \left((4\pi/\pi)\left(9^2/8^2\right)\right)^{1/3} = (4*81/64)^{1/3} = 1.717 \text{ ft}$$

and

$$r = ((V/\pi)(C_2/C_1))^{1/3} = ((4\pi/\pi)(8/9))^{1/3} = (4*8/9)^{1/3} = 1.526 \text{ ft}.$$

Now the height is larger than the radius.

$$V = \pi r^2 h = 3.1416 * 1.717 \text{ ft} * (1.526 \text{ ft})^2 = 12.6 \text{ ft}^3$$

and

$$Y = C_1 \pi r^2 + C_2 2\pi rh = 9*3.14*1.526^2 + 8*2*3.14*1.526*1.717$$
$$= \$65.8 + \$131.6$$
$$= \$197.4.$$

Note that the cost of the second term is still twice that of the first term even though only the cost for the sides was changed. This is not what one would expect if looking only at the primal objective equation as only C_2 was reduced and C_1 was the same.

5.3   EVALUATIVE QUESTIONS

1. A trash can is designed to hold 3 cubic meters of trash. Determine the cost and the design parameters (radius and height) in meters for if the costs C_1 and C_2 are 20 Swedish Kroner per square meter and 10 Swedish Kroner per square meter, respectively.

2. If the volume is doubled to 6 cubic meters, what are the new dimensions and cost?

3. If the unit costs are the same, both being 20 Swedish Kroner per square meter, what are the dimensions and the total cost and the two cost components?

4. If the trash can is to have a lid which will have the same diameter as the bottom of the trash can, what are the cost and dimensions of the trash can with the lid?

5.4   REFERENCES

[1] Bjorn Olaf Jonsson, Project Report for Advanced Manufacturing Processes, Spring Semester 2009. Industrial and Management Systems Engineering, West Virginia University.

CHAPTER 6

The Building Area Design Case Study

6.1 INTRODUCTION

This problem comes from a paper presented by Marvin Gates and Amerigo Scarpa in a 1982 article in *Cost Engineering* [1]. An assumption was made that the volume was a constant, and the results obtained are thus different. They assumed that the slab on grade and two suspended slabs was two floors and in the geometric programming model it was taken as three floors. They solved the problem by taking derivatives, and the design relationships are not the same as those obtained by geometric programming, but the results for the initial problem are identical.

6.2 THE BUILDING AREA DESIGN PROBLEM

A new building is to be constructed with a specified volume of 8,000 cubic meters. The problem is to determine what the area and total height dimensions of the building should be. The length, width, and height of the building are variable and the number of floors is specified to be three, that is the ground floor (slab-on-grade) and two raised floors (suspended slab). The dimensions of the building and the costs are:

Dimensions

H = Height of building, meters = ?

L = Length of building, meters = ?

W = Width of Building, meters = ?

$V = L \times W \times H$ = Volume of Building = 8,000 m^3

(Minimum Volume Required).

Since there are three floors, the gross height of each floor level would be $H/3$.

Cost Terms

N = Number of suspended slabs = 2

C_f = Cost of suspended slab, $/m^2 = $100/m^2

C_r = Cost of roof, $/m^2 = $70/m^2

C_s = Cost of slab-on-grade, $/m^2 = $30/m^2

C_w = Cost of walls, $/m^2 = $100/m^2.

The total cost is given by:

C (total) = Cost of Slab on Grade + Suspended Slab Cost + Roof Cost + Wall Cost

$$C \text{ (total)} = C_s * W * L + N * C_f * W * L + C_r * W * L$$
$$+ (N + 1) * C_w * V * (1/W + 1/L). \qquad (6.1)$$

The problem is to minimize the cost to obtain the minimum volume required.
Solve the problem by geometric programming:

a) Determine design relationships for $H, L,$ and D in terms of the cost terms and V

b) Use the general design relationships to determine the numerical values of $L, H,$ and W.

c) Show that the primal and dual objective functions are equivalent.

6.3   PROBLEM SOLUTION

The primal objective function and constraint can be expressed as:

$$C \text{ (total)} = C_s * W * L + N * C_f * W * L + C_r * W * L$$
$$+ (N + 1) * C_w * V * (1/W + 1/L)$$
$$= (W * L) * (C_s + NC_f + C_r) + H * (W + L)3C_w$$
$$Y = C_{01} * W * L + C_{02} * H * W + C_{03} * H * L, \qquad (6.2)$$

where

$$C_{01} = (C_s + NC_f + C_r) = (30 + 2 * 100 + 70) = 300 \qquad (6.3)$$
$$C_{02} = 3C_w = 3 * 100 = 300 \qquad (6.4)$$
$$C_{03} = 3C_w = 3 * 100 = 300. \qquad (6.5)$$

Subject to

$$LWH >= V_{min} \quad \text{or}$$
$$V_{min}/(L * W * H) <= 1 \quad \text{or}$$
$$C_{11}/L * W * H <= 1 \qquad (6.6)$$

and

$$C_{11} = 8000.$$

From the coefficients and signs, the signum values for the dual would be:

$$\sigma_{01} = 1$$
$$\sigma_{02} = 1$$
$$\sigma_{03} = 1$$
$$\sigma_{11} = 1.$$

The Dual Formulation would be

Obj	$\omega_{01} + \omega_{02} + \omega_{03}$ $= 1$		(6.7)
W terms	$\omega_{01} + \omega_{02}$ $- \omega_{11} = 0$		(6.8)
H terms	$+ \omega_{02} + \omega_{03} - \omega_{11} = 0$		(6.9)
L terms	ω_{01} $+ \omega_{03} - \omega_{11} = 0.$		(6.10)

The degrees of difficulty are

$$D = T - (N + 1) = 4 - (3 + 1) = 0. \tag{6.11}$$

Since there are 4 variables and 4 equations, the degrees of difficulty are zero. Using these equations, the values of the dual variables are found to be:

$$\omega_{11} = 2/3$$
$$\omega_{03} = 1 - \omega_{11} = 1 - 2/3 = 1/3$$
$$\omega_{01} = 1 - \omega_{11} = 1/3$$
$$\omega_{02} = 1 - \omega_{11} = 1/3$$

and by definition

$$\omega_{00} = 1.$$

The dual objective function would be,

$$Y = d(\omega) = \sigma \left[\prod_{m=0}^{M} \prod_{t=1}^{T_m} (C_{mt}\omega_{mo}/\omega_{mt})^{\sigma_{mt}\omega_{mt}} \right]^{\sigma} \tag{6.12}$$

$$Y = \sigma_{00} [(C_{01} * \omega_{00}/\omega_{01})^{\sigma_{01}\omega_{01}} * (C_{02} * \omega_{00}/\omega_{02})^{\sigma_{02}\omega_{02}}$$
$$*(C_{03} * \omega_{00}/\omega_{03})^{\sigma_{03}\omega_{03}} * (C_{11} * \omega_{10}/\omega_{11})^{\sigma_{11}\omega_{11}}]^{\sigma_{00}} \tag{6.13}$$

$$= 1(300 * 1/(1/3))^{1 \times 1/3} * (300 * 1/(1/3))^{1 \times 1/3}$$
$$* (300 * 1/(1/3))^{1 \times 1/3} * (8000 * (2/3)/(2/3))^{1 \times 2/3}$$

$$= 900 * 8000^{2/3}$$

$$= 360,000. \tag{6.14}$$

Using the Primal-Dual relationships

$$C_{01}WL = \omega_{01}Y \tag{6.15}$$
$$C_{02}HW = \omega_{02}Y. \tag{6.16}$$

Dividing these results in

$$L/H = (C_{02}/C_{01})(\omega_{01}/\omega_{02}) = C_{02}/C_{01}$$

or

$$L = H(C_{02}/C_{01}). \tag{6.17}$$

Similarly,

$$C_{01}WL = \omega_{01}Y \tag{6.15}$$
$$C_{03}HL = \omega_{03}Y. \tag{6.18}$$

Dividing these results in

$$W/H = (C_{03}/C_{01})(\omega_{01}/\omega_{03}) = C_{03}/C_{01}$$

or

$$W = H(C_{03}/C_{01}). \tag{6.19}$$

From the volume equation,

$$V = L * W * H = H * (C_{02}/C_{01}) * H * (C_{03}/C_{01}) * H$$
$$= H^3 * (C_{02}/C_{01}) * (C_{03}/C_{01})$$

one obtains

$$H = \left[V * C_{01}^2/(C_{02} * C_{03})\right]^{1/3} \tag{6.20}$$

$$H = \left[8000 * 300^2/(300 * 300)\right]^{1/3} = 8000^{1/3} = 20$$

$$L = H * (C_{02}/C_{01}) = \left[V * C_{01}^2/(C_{02} * C_{03})\right]^{1/3} (C_{02}/C_{01})$$

$$= \left[V * C_{02}^2/(C_{01} * C_{03})\right]^{1/3} \tag{6.21}$$

$$= \left[8000 * 300^2/(300 * 300)\right]^{1/3} = 20$$

$$W = H * (C_{03}/C_{01}) = \left[V * C_{01}^2/(C_{02} * C_{03})\right]^{1/3} (C_{03}/C_{01})$$

$$= \left[V * C_{03}^2/(C_{01} * C_{03})\right]^{1/3} \tag{6.22}$$

$$= \left[8000 * 300^2/(300 * 300)\right]^{1/3} = 20.$$

The total height is the height of the building, and the gross height of each floor level would be $H/(N+1)$ or 6.67 m. Note that N is not in the solution directly, but it is included in the constant C_{01}. These values of L, W, and H are the same as those of the original paper [1]. Now that the primal variables of H, W, and L have been determined, one can evaluate the primal objective function using Equation (6.2).

$$Y = C_{01} * W * L + C_{02} * H * W + C_{03} * H * L \tag{6.2}$$

$$= 300 * 20 * 20 + 300 * 20 * 20 + 300 * 20 * 20$$

$$= 120,000 + 120,000 + 120,000$$

$$= 360,000. \tag{6.23}$$

Note that each of the primary terms of the objective contribute equally as expressed by the dual variables for the dual objective function and that the dual objective function and primal objective function are equivalent. These are the same results which were obtained by the authors in Reference [1].

6.4   MODIFIED BUILDING AREA DESIGN PROBLEM

Since all three constants in the primal solution were equivalent, a modified example will be considered where these cost values are different. Let the cost terms be:

N = Number of suspended slabs = 2

C_f = Cost of suspended slab, \$/m² = \$200/m²

C_r = Cost of roof, \$/m² = \$120/m²

C_s = Cost of slab-on-grade, \$/m² = \$80/m²

C_w = Cost of walls, \$/m² = \$100/m²

V = Building Volume = 15,625 m³.

The cost terms would be:

$$C_{01} = (C_s + NC_f + C_r) = (80 + 2 * 200 + 120) = 600$$
$$C_{02} = 3C_w = 3 * 100 = 300$$
$$C_{03} = 3C_w = 3 * 100 = 300$$
$$C_{11} = 15,625.$$

The dual variables are the same, so the dual objective function would be:

$$Y = \sigma_{00} \left[(C_{01} * \omega_{00}/\omega_{01})^{\sigma_{01}\omega_{01}} * (C_{02} * \omega_{00}/\omega_{02})^{\sigma_{02}\omega_{02}} \right.$$
$$\left. *(C_{03} \times \omega_{00}/\omega_{03})^{\sigma_{03}\omega_{03}} * (C_{11} * \omega_{10}/\omega_{11})^{\sigma_{11}\omega_{11}} \right]^{\sigma_{00}} \quad (6.13)$$
$$= 1(600 * 1/(1/3))^{1\times 1/3} * (300 * 1/(1/3))^{1\times 1/3}$$
$$* (300 * 1/(1/3))^{1\times 1/3} * (15,625 * (2/3)/(2/3))^{1\times 2/3}$$
$$= 12.1644 * 9.6549 * 9.6549 * 625$$
$$= 708,706. \quad (6.14)$$

The primal variables and primal solution would be

$$H = \left[V * C_{01}^2/(C_{02} * C_{03}) \right]^{1/3} \quad (6.20)$$
$$H = \left[15,625 * 600^2/(300 * 300) \right]^{1/3} = 39.684$$
$$L = H * (C_{02}/C_{01}) = \left[V * C_{01}^2/(C_{02} * C_{03}) \right]^{1/3} (C_{02}/C_{01})$$
$$L = \left[V * C_{02}^2/(C_{01} * C_{03}) \right]^{1/3} \quad (6.21)$$
$$L = \left[15.625 * 300^2/(600 * 300) \right]^{1/3} = 19.843$$
$$W = H * (C_{03}/C_{01}) = \left[V * C_{01}^2/(C_{02} * C_{03}) \right]^{1/3} (C_{03}/C_{01})$$
$$W = \left[V * C_{03}^2/(C_{01} * C_{03}) \right]^{1/3} \quad (6.22)$$
$$W = \left[15.625 * 300^2/(600 * 300) \right]^{1/3} = 19.843.$$

Equations (6.20)–(6.22) are considered to be the design equations for H, L, and W. Note that the volume meets the required minimum as indicated by:

$$V = 39.684 * 19,843 * 19.843 = 15,625.$$

The primal objective function would be:

$$Y = C_{01} * W * L + C_{02} * H * W + C_{03} * H * L$$
$$= 600 * 19.843 * 19.843 + 300 * 39.684 * 19.843 + 300 * 39.684 * 19.843$$
$$= 236,247 + 236,235 + 236,235 = 708,717.$$

These values indicate that the contributions of each of the three terms of the primal are equivalent as indicated by the dual variables even though the constants have different values. The values also indicate that the building is a square shape which is also the same conclusion with the Optimal Box Design Case Study in Chapter 4. Other constraints can result in cases where the length and width would not be the same.

6.5   FIXED ROOM HEIGHT AREA DESIGN PROBLEM

In the initial problem presented, the height was a variable and resulted in a height of 20 m. For an office building, the room height would be approximately 2.5 meters; so how does this change the problem? The cost terms become:

N = Number of suspended slabs = 2

C_f = Cost of suspended slab, $/m^2$ = $100/m^2$

C_r = Cost of roof, $/m^2$ = $70/m^2$

C_s = Cost of slab-on-grade, $/m^2$ = $30/m^2$

C_w = Cost of walls, $/m^2$ = $100/m^2$.

The primal objective function and constraint can be expressed as:

$$C(\text{total}) = C_s * W * L + N * C_f * W * L + C_r * W * L + (N + 1)$$
$$* C_w * 2.5 * 2 * (L + W) \tag{6.24}$$
$$= (W * L) * (C_s + NC_f + C_r) + 3C_w * 2.5 * 2(L + W)$$
$$= (W * L) * (30 + 2 * 100 + 70) + 3 * 100 * 2.5 * 2(L + W)$$
$$= (W * L) * (300) + 1500 * L + 1500 * W, \tag{6.25}$$

or in geometric programming form for the primal:

$$Y = C_{01} * W * L + C_{02} * L + C_{03} * W. \tag{6.26}$$

Subject to

$$LWH >= V_{\min} = 8000$$
$$8000/(L * W * 2.5) <= 1$$

or

$$C_{11}/L * W <= 1 \tag{6.27}$$

and

$$C_{11} = 3200.$$

From the coefficients and signs, the signum values for the dual would be:

$$\sigma_{01} = 1$$
$$\sigma_{02} = 1$$
$$\sigma_{03} = 1$$
$$\sigma_{11} = 1.$$

The Dual Formulation would be

Obj	$\omega_{01} + \omega_{02} + \omega_{03} \qquad = 1$	(6.28)
L terms	$\omega_{01} + \omega_{02} \qquad - \omega_{11} = 0$	(6.29)
W terms	$\omega_{01} \qquad + \omega_{03} - \omega_{11} = 0.$	(6.30)

The degrees of difficulty are

$$D = T - (N + 1) = 4 - (2 + 1) = 1. \tag{6.31}$$

This would normally cause a problem, but if one closely examines Equations (6.29) and (6.30), one observes that ω_{02} and ω_{03} are equal, and since C_{02} and C_{03} are also equal, the Primal-Dual Equations indicate that L and W must also be equal. That is:

$$C_{02}L = \omega_{02}Y \tag{6.32}$$
$$C_{03}W = \omega_{03}Y. \tag{6.33}$$

Using Equation (6.27), the values of L and W must both be 56.6, and $L * W$ would be 3,200, and the total cost would be:

$$
\begin{aligned}
C(\text{total}) &= (W * L) * (300) + 1,500 * L + 1,500 * W \\
&= 3,200 * 300 + 1,500 * 56.6 + 1,500 * 56.6 \\
&= 1,129,800.
\end{aligned}
\tag{6.34}
$$

The restriction of the room height from 6.67 m to 2.5 m greatly increased the area and cost, but the optimal shape is still a square shape. Note that when only the cost values changed, the formulas for the primal variables did not change, but the values of the variables did change as the cost terms changed. When the exponents of the primal change, the formulas do change.

6.6   EVALUATIVE QUESTIONS

1. Change the cost of the walls from \$150/m² to \$500/m² in the modified example problem in Section 6.4 and determine the values of the primal variables, primal objective function, and dual objective function.

2. If the volume of Problem 1 is doubled, what is the change in each of the dimensions $-L, W$, and H?

3. What is the area/perimeter ratio for the following shapes:

    a) Square $(L * L)$

    b) Rectangle with width twice the length $(L * 2L)$

    c) Rectangle with width half the length $(L * 0.5L)$

    d) Which of a), b), and c) is the shape with the lowest ratio of area/perimeter?

    e) What is the area/perimeter ratio for a circle of diameter d)?

6.7 REFERENCES

[1] Marvin Gates and Amerigo Scarpa, Optimum configuration of buildings and site utilization, *Cost Engineering*, Vol. 24, No. 6, pp. 333–346, December 1982. 29, 33

CHAPTER 7

The Open Cargo Shipping Box Case Study

7.1 PROBLEM STATEMENT AND GENERAL SOLUTION

The Open Cargo Shipping Box Case Study is the classic geometric programming problem, as it was the first illustrative problem presented in the first book [1] on geometric programming. This is a classic problem and is presented in more detail to emphasize the benefits of geometric programming with zero degrees of difficulty.

This problem presented here is expanded, as not only is the minimum total cost required, but it also determines the dimensions of the box. The modified problem is: Suppose that 400 cubic yards (V) of gravel must be ferried across a river. The gravel is to be shipped in an open cargo box of length x_1, width x_2, and height x_3. The sides and bottom of the box cost \$10 per square yard (A_1) and the ends of the box cost \$20 per square yard (A_2). The cargo box will have no salvage value and each round trip of the box on the ferry will cost 10 cents (A_3).

a) What is the minimum total cost of transporting the 400 cubic yards of gravel?

b) What are the dimensions of the cargo box?

c) What is the number of ferry trips to transport the 400 cubic yards of gravel?

Figure 7.1 illustrates the parameters of the open cargo shipping box.

The first issue is to determine the various cost terms to find the objective function. The ferry transportation cost can be determined by:

$$T1 = V * A_3/(x_1 * x_2 * x_3) = 400 * 0.10/(x_1 * x_2 * x_3) = 40/(x_1 * x_2 * x_3). \qquad (7.1)$$

The cost for the ends of the box (2 ends) is determined by:

$$T2 = 2 * (x_2 * x_3) * A_2 = 2 * (x_2 * x_3) * 20 = 40 * (x_2 * x_3). \qquad (7.2)$$

The cost for the sides of the box (2 sides) is determined by:

$$T3 = 2 * (x_1 * x_3) * A_1 = 2 * (x_1 * x_3) * 10 = 20 * (x_1 * x_3). \qquad (7.3)$$

The cost for the bottom of the box is determined by:

$$T4 = (x_1 * x_2) * A_1 = (x_1 * x_2) * 10 = 10 * (x_1 * x_2). \qquad (7.4)$$

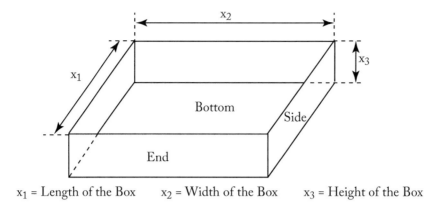

x_1 = Length of the Box x_2 = Width of the Box x_3 = Height of the Box

Figure 7.1: Open cargo shipping box.

The objective function (Y) is the sum of the four terms and is:

$$Y = T1 + T2 + T3 + T4 \tag{7.5}$$

$$Y = 40/(x_1 * x_2 * x_3) + 40 * (x_2 * x_3) + 20 * (x_1 * x_3) + 10 * (x_1 * x_2). \tag{7.6}$$

The primal objective function can be written in terms of generic constants for the cost variables to obtain a generalized solution.

$$Y = C_1/(x_1 * x_2 * x_3) + C_2 * (x_2 * x_3) + C_3 * (x_1 * x_3) + C_4 * (x_1 * x_2). \tag{7.7}$$

where $C_1 = 40, C_2 = 40, C_3 = 20$ and $C_4 = 10$.

From the coefficients and signs, the signum values for the dual are:

$$\sigma_{01} = 1$$
$$\sigma_{02} = 1$$
$$\sigma_{03} = 1$$
$$\sigma_{04} = 1.$$

The dual formulation is:

$$\text{Objective Function} \quad \omega_{01} + \omega_{02} + \omega_{03} + \omega_{04} = 1 \tag{7.8}$$

$$x_1 \text{terms} \qquad -\omega_{01} \qquad +\omega_{03} + \omega_{04} = 0 \tag{7.9}$$

$$x_2 \text{terms} \qquad -\omega_{01} + \omega_{02} \qquad + \omega_{04} = 0 \tag{7.10}$$

$$x_3 \text{terms} \qquad -\omega_{01} + \omega_{02} + \omega_{03} \qquad = 0. \tag{7.11}$$

Using these equations, the values of the dual variables are found to be:

$$\omega_{01} = 2/5$$
$$\omega_{02} = 1/5$$
$$\omega_{03} = 1/5$$
$$\omega_{04} = 1/5$$

and by definition

$$\omega_{00} = 1.$$

Thus, the dual variables indicate that the first term of the primal expression is twice as important as the other three terms. The degrees of difficulty are equal to:

$$D = T - (N + 1) = 4 - (3 + 1) = 0. \tag{7.12}$$

The objective function can be found using the dual expression:

$$Y = d(\omega) = \sigma \left[\prod_{m=0}^{M} \prod_{t=1}^{T_m} (C_{mt}\omega_{mo}/\omega_{mt})^{\sigma_{mt}\omega_{mt}} \right]^{\sigma} \tag{7.13}$$

$$= 1 \left[\left[\{(C_1 * 1/(2/5))\}^{(1*2/5)} \right] \left[\{C_2 * 1/(1/5))\}^{(1*1/5)} \right] \right.$$

$$\left. \left[\{C_3 * 1/(1/5)\}^{(1*1/5)} \right] \left[\{C_4 * 1/(1/5)\}^{(1*1/5)} \right] \right]^{1}$$

$$= 100^{2/5} * 200^{1/5} * 100^{1/5} * 50^{1/5}$$

$$= 100^{2/5} * 1000000^{1/5}$$

$$= 100^{2/5} * 100^{3/5}$$

$$= \$100.$$

Thus, the minimum cost for transporting the 400 cubic yards of gravel across the river is $100. The design equations for the primal variables can be determined from the relationships between the primal and dual as:

$$C_1/(x_1 * x_2 * x_3) = \omega_{01}Y = (2/5)Y \tag{7.14}$$
$$C_2 * x_2 * x_3 = \omega_{02}Y = (1/5)Y \tag{7.15}$$
$$C_3 * x_1 * x_3 = \omega_{03}Y = (1/5)Y \tag{7.16}$$
$$C_4 * x_1 * x_2 = \omega_{04}Y = (1/5)Y. \tag{7.17}$$

If one combines Equations (7.15) and (7.16) one can obtain the relationship:

$$x_2 = x_1 * (C_3/C_2). \tag{7.18}$$

If one combines Equations (7.16), (7.17), and (7.18), one can obtain the relationship:

$$x_3 = x_2 * (C_4/C_3) = x_1 * (C_3/C_2) * (C_4/C_3) = x_1 * (C_4/C_2). \tag{7.19}$$

If one combines Equations (7.14) and (7.15), one can obtain the relationship:

$$x_1 * x_2^2 * x_3^2 = (1/2) * (C_1/C_2). \tag{7.20}$$

Using the values for x_2 and x_3 in Equation (7.20), one can obtain:

$$x_1 = \left[(1/2) * \left(C_1 C_2^3 / \left(C_3^2 C_4^2\right)\right)\right]^{1/5}. \tag{7.21}$$

Similarly, one can solve for x_2 and x_3 and the equations would be:

$$x_2 = \left[(1/2) * \left(C_1 C_3^3 / \left(C_2^2 C_4^2\right)\right)\right]^{1/5} \tag{7.22}$$

and

$$x_3 = \left[(1/2) * \left(C_1 C_4^3 / \left(C_2^2 C_3^2\right)\right)\right]^{1/5}. \tag{7.23}$$

Now using the values of $C_1 = 40, C_2 = 40, C_3 = 20$, and $C_4 = 10$, the values of x_1, x_2, and x_3 can be determined using the design Equations (7.21), (7.22), and (7.23) as:

$$x_1 = \left[(1/2) * \left(40 * 40^3 / \left(20^2 10^2\right)\right)\right]^{1/5} = [32]^{1/5} = 2 \text{ yards}$$
$$x_2 = \left[(1/2) * \left(40 * 20^3 / \left(40^2 10^2\right)\right)\right]^{1/5} = [1]^{1/5} = 1 \text{ yard}$$
$$x_3 = \left[(1/2) * \left(40 * 10^3 / \left(40^2 20^2\right)\right)\right]^{1/5} = [0.03125]^{1/5} = 0.5 \text{ yard}.$$

Thus, the box is 2 yards in length, 1 yard in width, and 0.5 yard in height. The total box volume is the product of the three dimensions, which is 1 cubic yard.

The number of trips the ferry must make is 400 cubic yards/1 cubic yard/trip = 400 trips.

If one uses the primal variables in the primal equation, the values are:

$$Y = 40/(x_1 * x_2 * x_3) + 40 * (x_2 * x_3) + 20 * (x_1 * x_3) + 10 * (x_1 * x_2) \tag{7.6}$$
$$Y = 40/(2 * 1 * 1/2) + 40 * (1 * 1/2) + 20 * (2 * 1/2) + 10 * (2 * 1) \tag{7.24}$$
$$Y = 40 + 20 + 20 + 20$$
$$= \$100.$$

Note that the primal and dual give the same result for the objective function. Note that the terms of the primal solution (40, 20, 20, 20) are in the same ratio as the dual variables (2/5, 1/5, 1/5, 1/5). This ratio will remain constant even as the values of the constants change and this is important in the ability to determine which of the terms are dominant in the total cost. Thus, the transportation cost is twice the cost of the box bottom, and the box bottom is the same as the cost of the box sides and the same as the cost of the box ends. This indicates the optimal design relationships between the costs of the various box components and the transportation cost associated with the design. Although the design relationships are the same, the values of the design parameters can change significantly as the cost values change.

Let us consider a large increase in the transportation cost from 0.10 \$/trip to 1.00 \$/trip. This will change C_1 from \$40 to \$400 and the other cost coefficients terms will remain the same at 40, 20, and 10. The dual variables are independent of the cost coefficients, so they will remain the same as well as the design equations developed. Thus, using the design Equations of (7.21), (7.22), and (7.23) the new values would be:

$$x_1 = \left[(1/2) * \left(C_1 C_2^3 / \left(C_3^2 C_4^2\right)\right)\right]^{1/5}$$
$$= \left[(1/2) * \left(400 * 40^3 / \left(20^2 10^2\right)\right)\right]^{1/5} = [320]^{1/5} = 3.17 \text{ yards}$$
$$x_2 = \left[(1/2) * \left(C_1 C_3^3 / \left(C_2^2 C_4^2\right)\right)\right]^{1/5}$$
$$= \left[(1/2) * \left(400 * 20^3 / \left(40^2 10^2\right)\right)\right]^{1/5} = [10]^{1/5} = 1.58 \text{ yards}$$
$$x_3 = \left[(1/2) * \left(C_1 C_4^3 / \left(C_2^2 C_3^2\right)\right)\right]^{1/5}$$
$$= \left[(1/2) * \left(400 * 10^3 / \left(40^2 20^2\right)\right)\right]^{1/5} = [0.3125]^{1/5} = 0.792 \text{ yards.}$$

The box volume is the product of the three variables and is 3.97 (or 4 for rounding) cubic yards. The primal objective function is

$$Y = 400/(x_1 * x_2 * x_3) + 40 * (x_2 * x_3) + 20 * (x_1 * x_3) + 10 * (x_1 * x_2)$$
$$Y = 400/(3.17 * 1.58 * 0.792) + 40 * 1.58 * 0.792 + 20 * 3.17 * 0.792 + 10 * 3.17 * 1.58$$
$$Y = 100. + 50 + 50. + 50$$
$$Y = \$250.$$

Thus, although the transportation cost increased by a factor of 10, the total cost increased by a factor of 2.5 and the box size increased by a factor of 4 from 1 yard to 4 yards, and the number of trips decreased by a factor of 4. Note that the dimensions increased by the factor of $10^{0.2}$ or 1.58 for all three components. The dual objective function indicates the total cost would increase by a factor of $10^{0.4}$ or 2.51. Note that the ratio of the cost terms are the same as that of the initial solution, that is 2:1:1:1. This type of analysis is a major advantage of geometric programming with zero degrees of difficulty over all other methods of optimization.

7.2 EVALUATIVE QUESTIONS

1. The ferry cost for a round trip is increased from \$0.10 to \$3.20. What is the new total cost, the new box dimensions, and the number of ferry trips required to transport the 400 cubic yards of gravel?

2. The ferry cost for a round trip is increased from $3.20 to $213.06. What is the new total cost, the new box dimensions, and the number of ferry trips required to transport the 400 cubic yards of gravel?

3. A cover must be added to the box and it is made having the same costs as the box bottom. Determine the new total cost, the new box dimensions, and the number of ferry trips required to transport the 400 cubic yards of gravel.

7.3   REFERENCES

[1] R. J. Duffin, E. L. Peterson, and C. Zener, *Geometric Programming*, p. 6, John Wiley & Sons, NY, 1967. 39

CHAPTER 8

Metal Casting Cylindrical Side Riser Case Study

8.1 INTRODUCTION

The riser design problem in metal casting is always a concern for foundry engineers. The riser (also called feeders in many parts of the world) is an amount of additional metal added to a metal casting to move the thermal center of the casting and riser into the riser so there will be no solidification shrinkage in the casting. The risers are typically shaped as cylinders, as other shapes are difficult for the molding process and this shape has been successfully used for decades. The riser also has other design conditions, such as to supply sufficient feed metal, but thermal design issues are typically the primary concern. This is the topic where I first applied geometric programming [1, 2]. There are several additional papers [3–5] on riser design using geometric programming for side riser, top riser, insulated riser, and many other riser design types.

For a riser to be effective, the riser must solidify after the casting in order to provide liquid feed metal to the casting. The object is to have a riser of minimum volume to improve the yield of the casting process, which improves the economics of the process. The case study considered is for a cylindrical side riser, which consists of a cylinder of height H and diameter D. Figure 8.1 indicates the relationship between the casting and the side riser and the parameters of the riser.

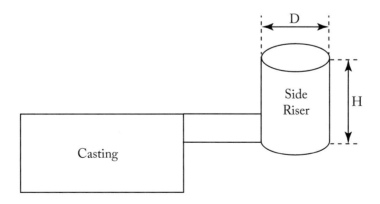

Figure 8.1: The cylindrical side riser.

The theoretical basis for riser design is Chvorinov's Rule, which is:

$$t = K(V/SA)^2, \tag{8.1}$$

where

t = solidification time (minutes or seconds)

K = solidification constant for molding material (minutes/in^2 or seconds/cm^2)

V = riser volume (in^3 or cm^3)

SA = cooling surface area of the riser (in^2 or cm^2).

The objective is to design the smallest riser such that

$$t_R \geq t_C, \tag{8.2}$$

where

t_R = solidification time of the riser

t_C = solidification time of the casting.

This constraint Equation (8.2) can be written as:

$$K_R(V_R/SA_R)^2 \geq K_C(V_C/SA_C)^2. \tag{8.3}$$

The riser and the casting are assumed to be molded in the same material so the K_R and K_C are equal and thus the equation can be written as:

$$(V_R/SA_R) \geq (V_C/SA_C). \tag{8.4}$$

The casting has a specified volume and surface area; the right-hand side of the equation can be expressed as a constant $Y = (V_C/SA_C)$, which is called the casting modulus (M_c), and Equation (8.4) becomes

$$(V_R/SA_R) \geq Y. \tag{8.5}$$

The volume and surface of the cylindrical riser can be written as:

$$V_R = \pi D^2 H/4 \tag{8.6}$$

$$SA_R = \pi DH + 2\pi D^2/4. \tag{8.7}$$

The surface area expression neglects the connection area between the casting and the riser as the effect is small. Thus, Equation (8.5) can be rewritten as:

$$\left(\pi D^2 H/4\right) / \left(\pi DH + 2\pi D^2/4\right) = (DH)/(4H + 2D) \geq Y. \tag{8.8}$$

The constraint must be rewritten in the less-than-equal form with the right-hand side being less than or equal to one, which becomes

$$4YD^{-1} + 2YH^{-1} \leq 1. \tag{8.9}$$

8.2   PROBLEM FORMULATION AND GENERAL SOLUTION

The primal form of the side cylindrical riser design problem can be stated as:

$$\text{Minimize:} \quad V = \pi D^2 H/4 \tag{8.10}$$

$$\text{Subject to:} \quad 4YD^{-1} + 2YH^{-1} \leq 1. \tag{8.11}$$

From the coefficients and signs, the signum values for the dual are:

$$\sigma_{01} = 1$$
$$\sigma_{11} = 1$$
$$\sigma_{12} = 1$$
$$\sigma_1 \ = 1.$$

The dual problem formulation is:

$$\text{Objective Function} \quad \omega_{01} \qquad\qquad = 1 \tag{8.12}$$

$$D \text{ terms} \quad 2\omega_{01} - \omega_{11} \quad\ = 0 \tag{8.13}$$

$$H \text{ terms} \quad \omega_{01} \qquad - \omega_{12} = 0. \tag{8.14}$$

The degrees of difficulty (D) are equal to:

$$D = T - (N + 1) = 3 - (2 + 1) = 0. \tag{8.15}$$

Using Equations (8.12) to (8.14), the values of the dual variables were found to be:

$$\omega_{01} = 1$$
$$\omega_{11} = 2$$
$$\omega_{12} = 1.$$

Using the linearity inequality equation,

$$\omega_{10} = \omega_{mt} = \sigma_m \sum \sigma_{mt}\omega_{mt}$$
$$= (1) * (1 * 2 + 1 * 1) = 3 > 0 \quad \text{where} \quad m = 1 \text{ and } t = 1. \tag{8.16}$$

The objective function can be found using the dual expression:

$$Y = d(\omega) = \sigma \left[\prod_{m=0}^{M} \prod_{t=1}^{T_m} (C_{mt}\omega_{mo}/\omega_{mt})^{\sigma_{mt}\omega_{mt}} \right]^{\sigma} \tag{8.17}$$

$$= 1 \left[\left[\{(\pi/4 * 1/1)\}^{(1*1)} \right] \left[\{(4Y * 3/2)\}^{(1*2)} \right] \left[\{(2Y * 3/1)\}^{(1*1)} \right] \right]^{1}$$
$$= (\pi/4) * (6Y)^2 * (6Y)$$
$$= (\pi/4) * (6Y)^3. \tag{8.18}$$

The values for the primal variables can be determined from the relationships between the primal and dual as:

$$4YD^{-1} = \omega_{11}/\omega_{10} = 2/3 \qquad (8.19)$$

and

$$2YH^{-1} = \omega_{11}/\omega_{10} = 1/3. \qquad (8.20)$$

The design equations for H and D can be determined as:

$$D = 6Y \qquad (8.21)$$

and

$$H = 6Y. \qquad (8.22)$$

Using (8.21) and (8.22) in Equation (8.6) for the primal, one obtains the design equation for the volume as:

$$V_R = (\pi/4) * (6Y)^3. \qquad (8.23)$$

Equations (8.18) and (8.23) are identical, which is what should happen, as the primal and dual objective functions must be identical. The values for the riser diameter and the riser height are both six times the casting modulus. These relationships hold for the side cylindrical riser design with negligible effects for the connecting area. This also indicates that the riser height and riser diameter are equal for the side riser. These equations were validated when using the data of another researcher [2] and comparing successful shape. Designs for other riser shapes and with insulating materials using geometric programming are given in the references [3–5].

8.3   CYLINDRICAL SIDE RISER EXAMPLE

A rectangular plate casting with dimensions $L = W = 10$ cm and $H = 4$ cm is to be produced, and a cylindrical side riser is to be used. The optimal dimensions for the side riser can be obtained from the casting modulus Y and Equations (8.21) and (8.22). The casting modulus is obtained by:

$$Y = (V_C/SA_C)$$
$$= (10 \text{ cm} \times 10 \text{ cm} \times 4 \text{ cm})/[2(10 \text{ cm} \times 10 \text{ cm}) + 2(10 \text{ cm} \times 4 \text{ cm}) + 2(10 \text{ cm} \times 4 \text{ cm})]$$
$$= 400 \text{ cm}^3/360 \text{ cm}^2 = 1.111 \text{ cm}.$$

Thus,

$$H = 6Y = 6 \times 1.111 \text{ cm} = 6.67 \text{ cm}$$
$$D = 6Y = 6 \times 1.111 \text{ cm} = 6.67 \text{ cm}.$$

The volume of the riser can be obtained from Equation (8.23) as:

$$V_R = (\pi/4)(6Y)^3 = 233 \text{ cm}^3.$$

The dual variables for the constraint indicate the relative amount of the surface area constraint where $\omega_{11} = 2$ represents the total surface area of the side, and $\omega_{12} = 1$ represents the total area of the top and bottom of the riser. This ratio was determined before the primal variables were determined, and using the primal variables the surfaces of the areas are:

$$A(\text{side}) = \pi DH = \pi 6.67 \times 6.67 = 139.8$$

$$A(\text{top \& bottom}) = 2 \times \pi D^2/4 = 2 \times \pi \times 6.67^2/4 = 69.9.$$

Thus, once the modulus of the casting is determined, the riser height, diameter, and volume can be determined using Equations (8.21)–(8.23).

8.4   EVALUATIVE QUESTIONS

1. A cylindrical side riser is to be designed for a rectangular metal casting (5 cm $\times$ 8 cm $\times$ 10 cm) which has a surface area of 340 cm^2 and a volume of 400 cm^3. The hot metal cost is 100 Rupees per kg and the metal density is 3.0 gm/cm^3.

   a) What are the dimensions in centimeters for the side riser (H and D)?

   b) What is the volume of the side riser (cm^3)?

   c) What is the metal cost of the side riser (Rupees)?

   d) What is the metal cost of the casting (Rupees)?

2. Instead of a side riser a top riser is to be used; that is, the riser is placed on the top surface of the casting. The cooling surface area for the top riser is:

$$SA_R = \pi DH + \pi D^2/4.$$

Show that for the top riser $D = 6Y$ and $H = 3Y$.

8.5   REFERENCES

[1] R. C. Creese, Optimal riser design by geometric programming, *AFS Cast Metals Research Journal*, Vol. 7, pp. 118–121, 1971. 45

[2] R. C. Creese, Dimensioning of risers for long freezing range alloys by geometric programming, *AFS Cast Metals Research Journal*, Vol. 7, pp. 182–184, 1971. 45, 48

[3] R. C. Creese, Generalized riser design by geometric programming, *AFS Transactions*, Vol. 87, pp. 661–664, 1979. 45, 48

[4] R. C. Creese, An evaluation of cylindrical riser designs with insulating materials, *AFS Transactions,* Vol. 87, pp. 665–669, 1979.

[5] R. C. Creese, Cylindrical top riser design relationships for evaluating insulating materials, *AFS Transactions,* Vol. 89, pp. 345–348, 1981. 45, 48

CHAPTER 9

Inventory Model Case Study

9.1 PROBLEM STATEMENT AND GENERAL SOLUTION

The basic inventory model is to minimize the sum of the unit set-up costs and the unit inventory holding costs. The objective is to determine the optimal production quantity which will minimize the total costs. The problem has been solved using the method of Lagrange Multipliers, but it can also easily be solved using geometric programming, which permits a general solution for the production quantity in terms of the constant parameters.

The assumptions for the model are:

1. Replenishment of the order is instantaneous

2. No shortage is permitted

3. The order quantity is a batch.

The model can be formulated in terms of annual costs as:

$$\text{Total Cost} = \text{Total Unit Costs} + \text{Annual Inventory Carrying Cost}$$
$$+ \text{Annual Set-up Cost} \tag{9.1}$$
$$TC = DC_u + CA + SD/Q, \tag{9.2}$$

where

$TC = $ Total Annual Cost (\$)
$D = $ Annual Demand (pieces/year)
$C_u = $ Item Unit Cost (\$/piece)
$C = $ Inventory Carrying Cost (\$/piece-yr)
$A = $ Average Inventory (Pieces)
$S = $ Set-up Cost (\$/set-up)
$Q = $ Order Quantity (Pieces/order).

The average inventory for this model is given by:

$$A = Q/2. \tag{9.3}$$

Thus, the model can be formulated as:

$$TC = DC_u + CQ/2 + SD/Q. \tag{9.4}$$

The model can be formulated in general terms as:

$$\text{Minimize} \quad \text{Cost}(Y) = C_{00} + C_{01}Q + C_{02}/Q, \tag{9.5}$$

where

$$C_{00} = DC_u$$
$$C_{01} = C/2$$
$$C_{03} = SD.$$

Since C_{00} is a constant, the objective function can be rewritten for the variable cost terms as this reduces the number of terms and thus the degrees of freedom by one and thus:

$$\text{Minimize} \quad \text{Variable Cost}(Y) = Y_{vc} = C_{01}Q + C_{02}/Q. \tag{9.6}$$

From the coefficients and signs, the signum values for the dual are:

$$\sigma_{01} = 1$$
$$\sigma_{02} = 1.$$

Thus, the dual formulation would be:

$$\text{Objective Function [using Equation (9.6)]} \quad \omega_{01} + \omega_{02} = 1 \tag{9.7}$$
$$Q \text{ terms [using Equation (9.6)]} \quad \omega_{01} - \omega_{02} = 0. \tag{9.8}$$

The degrees of difficulty are equal to:

$$D = T - (N + 1) = 2 - (1 + 1) = 0. \tag{9.9}$$

Thus, one has the same number of variables as equations, so this can be solved by simultaneous equations as these are linear equations. Using Equations (9.7) and (9.8), the values for the dual variables are found to be:

$$\omega_{01} = 1/2$$
$$\omega_{02} = 1/2$$

and by definition

$$\omega_{00} = 1.$$

The objective function can be found using the dual expression:

$$d(\omega) = \sigma \left[\prod_{m=0}^{M} \prod_{t=1}^{T_m} (C_{mt}\omega_{mo}/\omega_{mt})^{\sigma_{mt}\omega_{mt}} \right]^{\sigma} \tag{9.10}$$

$$d(\omega) = 1 \left[\{(C_{01} * 1)/(1/2)\}^{(1)*(1/2)} * \{(C_{02} * 1)/(1/2)\}^{(1)*(1/2)} \right]^{1}$$

$$= 1 \left[\{(2C_{01})^{1/2}\} * \{(2C_{02})^{1/2}\} \right]$$

$$= 2C_{01}^{1/2}C_{02}^{1/2}$$

$$= 2(C/2 * SD)^{1/2}$$

or

$$Y_{vc} = (2C * SD)^{1/2} \tag{9.11}$$

and

$$TC = DC_u + Y_{vc}$$

or

$$TC = D * C_u + (2C * SD)^{1/2}. \tag{9.12}$$

The solution has been determined without finding the value for Q. Also note that the dual expression and total cost expressions are expressed in constants and thus the answer can be found without having to resolve the entire problem, as one only needs to use the new constant values. Note the importance of removing the constant term in the original objective function and solving for the variable objective function as it reduced the degrees of difficulty and made the solution much easier.

9.2   INVENTORY EXAMPLE PROBLEM

The values for the parameters for the example problem are:

D = Annual Demand (pieces/year) = $100,000$/yr

C_u = Item Unit Cost ($/piece) = 1.5/piece

C = Inventory Carrying Cost ($/piece-yr) = 0.20/piece-year

A = Average Inventory (Pieces) = $Q/2$

S = Set-up Cost ($/set-up) = 400/set-up

Q = Order Quantity (Pieces/order) = Q.

Note the total cost can be found using Equation (9.12) as:

$$TC = 100,000 * 1.5 + (2 * 0.20 * 400 * 100,000)^{1/2} \tag{9.13}$$
$$= 150,000 + 4,000$$
$$= \$154,000. \tag{9.14}$$

The value of Q can be determined from the primal-dual relationships which are:

$$(C/2)Q = \omega_{01} Y_{vc} = (1/2)(2C * SD)^{1/2}$$

or

$$Q = (2SD/C)^{1/2}. \tag{9.15}$$

Equation (9.15) would be considered as the design equation for Q in terms of the input constants S, D, and C.

And for the example problem

$$Q = (2 * 400 * 100,000/0.20)^{1/2}$$
$$Q = 20,000 \text{ units.} \tag{9.16}$$

The primal problem can now be evaluated using Equation (9.4) as:

$$TC = DC_u + CQ/2 + SD/Q \tag{9.4}$$
$$TC = 100,000 * 1.5 + 0.20 * 20,000/2 + 400 * 100,000/20,000$$
$$= 150,000 + 2,000 + 2,000.$$
$$= \$154,000. \tag{9.17}$$

The number of set-ups per year would be 100,000/20,000 or five set-ups per year.

Thus, the primal and dual give identical values for the solution of the problem as indicated by Equations (9.14) and (9.17). The dual variables are both equal to 1/2, which implies that the total annual inventory carrying cost and the total annual set-up costs are equal, as illustrated by the example, and they should always be equal. If the inventory cost is reduced to 0.10$/pc-yr and the set-up cost increases to $800 and the demand remains constant at 100,000 units, the order quantity becomes:

$$Q = (2SD/C)^{1/2} = (2 * 800 * 100,000/0.10)^{1/2} = 40,000.$$

The total cost becomes

$$TC = DC_u + CQ/2 + SD/Q = 100,000 * 1.5 + 0.10 * 40,000/2 + 800 * 100,000/40,000$$
$$= 150,000 + 2,000 + 2,000 = 154,000.$$

Note that the Annual Inventory Carrying Cost and Annual Set-up Cost are equal, as indicated by the dual variables, and that these two costs total only $4,000 and the fixed cost was $150,000. The lot size increased from 20,000 to 40,000 because of the inventory and set-up unit cost changes. The total inventory cost and total set-up costs remained the same, as the inventory unit cost decreased while the set-up cost increased. If both unit costs increased, the total cost would have increased.

9.3   EVALUATIVE QUESTIONS

1. The following data was collected on a new pump.

   D = Annual Demand (pieces/year) = 200/yr
   C_u = Item Unit Cost ($/piece) = $300/piece
   C = Inventory Carrying Cost ($/piece-yr) = $20/piece-year
   S = Set-up Cost ($/set-up) = $500/set-up.

   a. Determine the total cost for the 200 pumps during the year.

   b. Determine the average total cost per pump.

   c. What is the number of set-ups per year?

   d. What is the total inventory carrying cost for the year?

2. The demand for the pumps increased dramatically to 3,000 because of the oil spill in the Gulf.

   D = Annual Demand (pieces/year) = 3,000/yr
   C_u = Item Unit Cost ($/piece) = $300/piece
   C = Inventory Carrying Cost ($/piece-yr) = $20/piece-year
   S = Set-up Cost ($/set-up) = $500/set-up.

   a. Determine the total cost for the 3,000 pumps during the year.

   b. Determine the average total cost per pump.

   c. What is the number of set-ups per year?

   d. What is the total inventory carrying cost for the year?

3. Resolve Problem 2 if the order must be completed in one year and adjust your answers so the number of set-ups would be an integer.

9.4   REFERENCES

[1] H. Jung and C. M. Klein, Optimal inventory policies under decreasing cost functions via geometric programming, *European Journal of Operational Research*, 132, pp. 628–642, 2001.

CHAPTER 10

Process Furnace Design Case Study

10.1 PROBLEM STATEMENT AND SOLUTION

An economic process model was developed [1, 2] for an industrial metallurgical application. The annual cost for a furnace operation in which the slag-metal reaction is a critical factor of the process was considered, and a modified version of the problem is presented and illustrated in Figure 10.1. The objective was to minimize the annual cost, and the primal equation representing the model was:

$$Y = C_1 / (L^2 * D * T^2) + C_2 * L * D + C_3 L * D * T^4. \tag{10.1}$$

The model was subject to the constraint that:

$$D \leq L.$$

The constraint must be set in geometric programming for which would be:

$$(D/L) \leq 1. \tag{10.2}$$

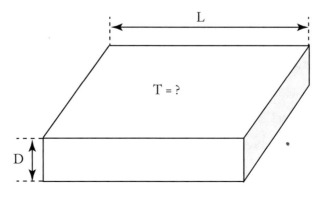

Figure 10.1: Process furnace.

where

D = Depth of the furnace (ft)
L = Characteristic Length of the furnace (ft)
T = Furnace Temperature (K)

The degrees of difficulty are:

$$D = T - (N + 1) = 4 - (3 + 1) = 0. \tag{10.3}$$

Thus, the problem has zero degrees of difficulty.

For the specific example problem, the values of the constants were:

$$C_1 = 10^{13} \left(\$ - \text{ft}^2 - \text{K}^2 \right)$$
$$C_2 = 100 \left(\$/\text{ft}^2 \right)$$
$$C_3 = 5 * 10^{-11} \left(\text{ft}^{-2} - \text{K}^{-4} \right).$$

From the coefficients and signs, the signum values for the dual are:

$$\sigma_{01} = 1$$
$$\sigma_{02} = 1$$
$$\sigma_{03} = 1$$
$$\sigma_{11} = 1$$
$$\sigma_1 = 1$$

The dual problem formulation is:

Objective Function	$\omega_{01} + \omega_{02} + \omega_{03}$	$= 1$	(10.4)
L terms	$-2\omega_{01} + \omega_{02} + \omega_{03} - \omega_{11} = 0$		(10.5)
D terms	$- \omega_{01} + \omega_{02} + \omega_{03} + \omega_{11} = 0$		(10.6)
T terms	$-2\omega_{01} \qquad +4\omega_{03} = 0.$		(10.7)

Using Equations (10.4) to (10.7), the values of the dual variables were found to be:

$$\omega_{01} = 0.4$$
$$\omega_{02} = 0.4$$
$$\omega_{03} = 0.2$$
$$\omega_{11} = -0.2.$$

The dual variables cannot be negative, and the negative value implies that the constraint is not binding, that is, it is a loose constraint. Thus, the problem must be reformulated without the constraint and the dual variable is forced to zero, that is, $\omega_{11} = 0$ and the equations resolved. This means that the constraint $D \leq L$ will be loose, that is, D will be less than L in the solution. Thus, one can remove the constraint and the primal becomes:

$$Y = C_1 / \left(L^2 * D * T^2 \right) + C_2 * L * D + C_3 L * D * T^4 \qquad (10.1)$$

and the constraint is removed from the problem.

The new dual becomes:

Objective Function	$\omega_{01} + \omega_{02} + \omega_{03} = 1$	(10.8)
L terms	$-2\omega_{01} + \omega_{02} + \omega_{03} = 0$	(10.9)
D terms	$-\omega_{01} + \omega_{02} + \omega_{03} = 0$	(10.10)
T terms	$-2\omega_{01} \qquad +4\omega_{03} = 0.$	(10.11)

Now the problem is that it has four equations to solve for three variables. The degrees of difficulty become:

$$D = D = T - (N + 1) = 3 - (3 + 1) = -1. \qquad (10.12)$$

When the degrees of difficulty become negative, it is best not to solve the problem but to reformulate it. One approach (not necessarily a good approach) would be to remove one of the equations and thus one of the primal variables.

If one examines Equations (10.9) and (10.10), one observes that Equation (10.9) is dominant over Equation (10.10) and thus Equation (10.10) will be removed from the dual formulation. This results in removing the variable D from the solution. The new dual formulation is:

Objective Function	$\omega_{01} + \omega_{02} + \omega_{03} = 1$	(10.13)
L terms	$-2\omega_{01} + \omega_{02} + \omega_{03} = 0$	(10.14)
T terms	$-2\omega_{01} \qquad +4\omega_{03} = 0.$	(10.15)

Note that the degrees of difficulty become zero for the dual, but the primal is still unchanged.

The new solution for the dual becomes:

$$\omega_{01} = 1/3$$
$$\omega_{02} = 1/2$$
$$\omega_{03} = 1/6$$

and by definition

$$\omega_{00} = 1.0.$$

The dual variables indicate that the second term is the most important, followed by the first term and then the third term.

The objective function can be found using the dual expression:

$$Y = d(\omega) = \sigma \left[\prod_{m=0}^{M} \prod_{t=1}^{T_m} (C_{mt}\omega_{mo}/\omega_{mt})^{\sigma_{mt}\omega_{mt}} \right]^{\sigma} \tag{10.16}$$

$$= 1 \left[\left[\{(C_1 * 1/(1/3)1)\}^{(1/3*1)} \right] \left[\{(C_2 * 1/(1/2))\}^{(1/2*1)} \right] \right.$$

$$\left. \left[\{(C_3/(1/6))\}^{(1/6*1)} \right] \right]^1$$

$$= 1 \left[\left[\{(1 * 10^{13} * 1/(1/3)1)\}^{(1/3*1)} \right] \left[\{(100 * 1/(1/2))\}^{(1/2*1)} \right] \right.$$

$$\left. \left[\{(5 * 10^{-11}/(1/6))\}^{(1/6*1)} \right] \right]^1$$

$$= \$11,370.$$

This can be expressed in a general form in terms of the constants as:

$$Y = (3C_1)^{1/3}(2C_2)^{1/2}(6C_3)^{1/6}. \tag{10.17}$$

The values for the primal variables can be determined from the relationships between the primal and dual which are:

$$C_1 * L^{-2} * D^{-1} * T^{-2} = \omega_{01} Y \tag{10.18}$$

$$C_2 * L * D = \omega_{02} Y \tag{10.19}$$

and

$$C_3 * L * D * T^4 = \omega_{03} Y. \tag{10.20}$$

The fully general expressions are somewhat difficult, but the variables can be expressed in terms of the constants and objective function as:

$$T = [(\omega_{03}/\omega_{02}) * (C_2/C_3)]^{1/4} \tag{10.21}$$

$$L = \left[\left(C_1 * (C_2 * C_3)^{(1/2)} \right) \right] / \left[\omega_{01} * (\omega_{02} * \omega_{03})^{(1/2)} * Y^2 \right] \tag{10.22}$$

$$D = \left[\left(\omega_{01} * \omega_{02}^{(3/2)} * \omega_{03}^{(1/2)} Y^3 \right) / \left(C_1 * C_2^{(3/2)} * C_3^{(1/2)} \right) \right]. \tag{10.23}$$

The expressions developed for the variables in terms of the constants in a reduced form were:

$$T = (C_2/3C_3)^{1/4} \tag{10.24}$$

$$L = (3C_1)^{1/3}(2C_2)^{-1/2}(6C_3)^{1/6} \tag{10.25}$$

$$D = 1. \tag{10.26}$$

Using the values of $Y = \$11{,}370$, $C_1 = 10^{13}$, $C_2 = 100, C_3 = 5 * 10^{-11}$, $\omega_{01} = 1/3$, $\omega_{02} = 1/2$ and $\omega_{03} = 1/6$, the values for the variables are:

$$T = 903 \text{ K}$$

$$L = 56.85 \text{ ft}$$

and

$$D = 1.00 \text{ ft.}$$

Using the values of the variables in the primal equation, the objective function is:

$$
\begin{aligned}
Y &= C_1 / \left(L^2 * D * T^2 \right) + C_2 * L * D + C_3 * L * D * T^4 \\
&= 10^{13} / \left(56.85^2 * 1 * 903^2 \right) + 100 * 56.85 * 1 + 5 * 10^{-11} * 56.85 * 1 * \left(903^4 \right) \\
&= 3{,}795 + 5{,}685 + 1{,}890 \\
&= \$11{,}370.
\end{aligned}
\tag{10.27}
$$

The values of the objective function for the primal and dual are identical, which implies that the values for the primal variables have been correctly obtained. The costs terms are in the same ratio as the dual variables; the third term is the smallest, the first term is twice the third term and the second term is three times the third term.

The constraint is loose, as $D = 1$ ft is much lower than $L = 56.85$ ft. The other item of interest was that equations dominated by other equations can prevent a solution and must be removed. The removal of the dominated equation was necessary to obtain a solution, but it is the cause of D being unity, as it was not considered as a variable in the dual. Even though the primal and dual are equivalent, this may **not** be a realistic solution to the problem. What happened is that when the dual constraint involving the depth D was removed, it forced the value of D to be unity to have a minimum effect on the solution. Similarly, removing the dual constraint involving the length L would result in forcing the value of L to be unity. However, all is not lost. The method to obtain a more valid solution is to reverse the constraint of the primal (as indicated in Evaluative Question 1) to be:

$$L \leq D. \tag{10.28}$$

The solution to this problem is left to the reader and it has positive dual variables.

Another approach would be to specify the depth as a specific value and a valid solution would be obtained. Since this problem is where the slag-metal interaction was important it would be similar to the open hearth furnace in steelmaking which had large surface area and the depth was much smaller than the length.

10.2 CONCLUSIONS

The highlights of this chapter are:

1. If a dual variable is negative, then that constraint is loose and should be removed. Although geometric programming formulates the constraints as inequalities, the solutions always tend to be at the boundary of the inequality.

2. When the degrees of difficulty are negative, the problem should be reformulated to a primal problem with zero or positive degrees of difficulty. Do not attempt to solve a problem with negative degrees of difficulty, and even though results may be obtained, they most likely would not be useful and the problem needs to be reformulated.

10.3   EVALUATIVE QUESTIONS

1. The problem constraint presented was given as $D \leq L$, but the designer decided that was incorrect and reversed the constraint to $L \leq D$. Resolve the problem and determine the dual and primal variables as well as the objective function.

2. Resolve the problem, making the initial assumption that $L = D$, and reformulate the primal and dual problems and find the variables and objective function.

3. Show that instead of eliminating the dual equation for the D terms, eliminate the dual equation for L terms in the dual and the value of L will become 1 in the solution.

4. Resolve the problem, but make the constraint $D \geq 5$. Note that the solution for the objective function will be $5^{1/3}$ times the value in Equation (10.25) for the problem solved and the values of D, L, and T will be different.

10.4   REFERENCES

[1] http://www.mpri.lsu/edu/textbook/Chapter3-b.htm (visited May 2009). 57

[2] W. H. Ray and J. Szekely, *Process Optimization with Applications in Metallurgy and Chemical Engineering*, John Wiley and Sons, Inc., NY (1973). 57

CHAPTER 11

The Gas Transmission Pipeline Case Study

11.1 PROBLEM STATEMENT AND SOLUTION

The energy crisis is with us today and one of the problems is in the transmission of energy. A gas transmission model was developed [1, 2] to minimize the total transmission cost of gas in a new gas transmission pipeline. The problem is more difficult than the previous case studies, as several of the exponents are constants but not integers. The primal expression for the cost developed was:

$$C = C_1 * L^{1/2} * V / \left(F^{0.387} * D^{2/3} \right) + C_2 * D * V + C_3/(L * F) + C_4 * F/L. \quad (11.1)$$

Subject to:

$$(V/L) \geq F.$$

The constraint must be restated in the geometric form as:

$$-(V/(LF)) \leq -1. \quad (11.2)$$

Where the variables are:

L = Pipe length between compressors (feet)

D = Diameter of Pipe (in)

V = Volume Flow Rate (ft^3/sec)

F = Compressor Pressure Ratio Factor.

The degrees of difficulty are equal to:

$$D = T - (N + 1) = 5 - (4 + 1) = 0. \quad (11.3)$$

Figure 11.1 is a sketch of the problem indicating the variables and is not drawn to scale. For the specific problem, the values of the constants were:

$$C_1 = 4.55 * 10^5$$
$$C_2 = 3.69 * 10^4$$
$$C_3 = 6.57 * 10^5$$
$$C_4 = 7.72 * 10^5.$$

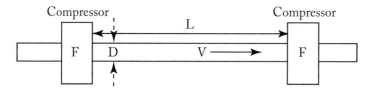

Figure 11.1: Gas transmission pipeline.

From the coefficients and signs, the signum values for the dual are:

$$\sigma_{01} = 1$$
$$\sigma_{02} = 1$$
$$\sigma_{03} = 1$$
$$\sigma_{04} = 1$$
$$\sigma_{11} = -1$$
$$\sigma_1 = -1.$$

The dual problem formulation is:

Objective Function	$\omega_{01} + \omega_{02} + \omega_{03} + \omega_{04} \qquad\qquad = 1$	(11.4)
L terms	$0.5\omega_{01} \qquad -\omega_{03} - \omega_{04} + \omega_{11} = 0$	(11.5)
F terms	$-0.387\omega_{01} \qquad -\omega_{03} + \omega_{04} + \omega_{11} = 0$	(11.6)
V terms	$\omega_{01} + \omega_{02} \qquad\qquad -\omega_{11} = 0$	(11.7)
D terms	$-0.667\omega_{01} + \omega_{02} \qquad\qquad = 0.$	(11.8)

Using Equations (11.4)–(11.8), the values of the dual variables were found to be:

$$\omega_{01} = 0.26087$$
$$\omega_{02} = 0.17391$$
$$\omega_{03} = 0.44952$$
$$\omega_{04} = 0.11570$$
$$\omega_{11} = 0.43478$$

and by definition

$$\omega_{00} = 1$$

and

$$\omega_{10} = \omega_{mt} = \sigma_m \sum \sigma_{mt}\omega_{mt}$$
$$= (-1) * (-1 * 0.43478) = 0.43478 \quad \text{where } m = 1 \text{ and } t = 1. \tag{11.9}$$

The objective function can be found using the dual expression:

$$Y = d(\omega) = \sigma \left[\prod_{m=0}^{M} \prod_{t=1}^{T_m} (C_{mt}\omega_{mo}/\omega_{mt})^{\sigma_{mt}\omega_{mt}} \right]^{\sigma} \tag{11.10}$$

$$= 1 \left[\left[\{(4.55 * 10^5 * 1/0.26087)\}^{(1*0.26087)} \right] \right.$$

$$* \left[\{(3.69 * 10^4 * 1/0.17391)\}^{(1*0.17391)} \right]$$

$$* \left[\{(6.57 * 10^5 * 1/0.44952)\}^{(1*0.22952)} \right]$$

$$* \left[\{(7.72 * 10^5 * 1/0.11570)\}^{(1*0.11570)} \right]$$

$$\left. * \left[\{(1 * 0.43478/0.43478)\}^{(-1*0.43478)} \right] \right]^{1}$$

$$= \$1,304,400 \quad \text{or} \quad \$1.3043 * 10^6/\text{yr}.$$

The values for the primal variables can be determined from the relationships between the primal and dual which are:

$$C_1 * L^{1/2} * V/(F^{0.387} * D^{2/3}) = \omega_{01}Y \tag{11.11}$$

$$C_2 * D * V = \omega_{02}Y \tag{11.12}$$

$$C_3/(L * F) = \omega_{03}Y \tag{11.13}$$

$$C_4 * F/L = \omega_{04}Y \tag{11.14}$$

$$V/(F * L) = \omega_{11}/\omega_{10} = 1. \tag{11.15}$$

The fully general expressions are somewhat difficult, but the variables can be expressed in terms of the constants and objective function as:

$$F = [(C_3\omega_{04})/(C_4\omega_{03})]^{1/2} \tag{11.16}$$

$$V = C_3/(\omega_{03} * Y) \tag{11.17}$$

$$L = [(C_3C_4)/(\omega_{03}\omega_{04})]^{1/2}/Y \tag{11.18}$$

$$D = [(\omega_{02} * \omega_{03})/(C_2 * C_3)] * Y^2. \tag{11.19}$$

Using the values of $Y = 1.3043 * 10^6$, $C_1 = 4.55 * 10^5$, $C_2 = 3.69 * 10^4$, $C_3 = 6.57 * 10^5$, $C_4 = 7.72 * 10^5$, $\omega_{01} = 0.26087$, $\omega_{02} = 0.17391$, $\omega_{03} = 0.44952$, $\omega_{04} = 0.11570$, and $\omega_{11} =$

0.43478 one obtains:

$$F = [(C_3 * \omega_{04})/(C_4 * \omega_{03})]^{1/2}$$
$$= [(6.47 * 10^5 * 0.11570)/(7.72 * 10^5 * 0.44952)]^{1/2} = 0.468$$
$$V = C_3/(\omega_{03} * Y)$$
$$= 6.57 * 10^5/(0.44952 * 1.3043 * 10^6) = 1.1205 \text{ ft}^3/\text{sec}$$
$$L = [(C_3 * C_4)/(\omega_{03} * \omega_{04})]^{1/2}/Y$$
$$= [(6.57 * 10^5 * 7.72 * 10^5)/(0.44952 * 0.1157)]^{1/2}/1.3043 * 10^6 = 2.3943 \text{ ft}$$
$$D = [(\omega_{02} * \omega_{03})/(C_2 * C_3)] * Y^2$$
$$= [(0.17391 * 0.44952)/(3.69 * 10^4 * 6.57 * 10^5)] * (1.3043 * 10^6)^2 = 5.4857 \text{ in.}$$

The primal expression can now be solved using the primal variables, and the contribution of each of the terms can be observed.

$$C = C_1 * L^{1/2} * V/\left(F^{0.387} * D^{2/3}\right) + C_2 * D * V + C_3/(L * F) + C_4 * F/L$$
$$= 4.55 * 10^5 * 2.3943^{1/2} * 1.1205/\left(0.468^{0.387} * 5.4857^{2/3}\right) + 3.69 * 10^4 * 5.4857 * 1.1205$$
$$+ 6.57 * 10^5/(2.3943 * 0.468) + 7.72 * 10^5 * 0.468/2.3943$$
$$= 3.4029 * 10^5 + 2.26814 * 10^5 + 5.8633 * 10^5 + 1.5090 * 10^5$$
$$= \$1,304,300.$$

The third term is slightly higher than the others, but all terms are of the same magnitude. Since the constraint is binding, that is, $V = L * F$ and the results indicate that the binding constraint holds as:

$$1.1205 = 2.3943 * 0.468 = 1.1205.$$

The values of the dual variables were more complex for this problem than the previous problems, but the values of these dual variables still have the same relationship to the terms of the primal cost function. The first dual variable, ω_{01}, was 0.26087, and the relation between the first cost term of the primal to the total cost is $3.4029*10^5/1.3043*10^6 = 0.2609$. The reader should show that the other dual variables have the same relationships between the terms of the primal cost function and the total primal cost.

11.2   EVALUATIVE QUESTIONS

1. Resolve the problem with the values of $C_1 = 6 * 10^5$, $C_2 = 5 * 10^4$, $C_3 = 7 * 10^5$, and $C_4 = 8 * 10^5$. Determine the effect upon the dual variables, the objective function, and the primal variables. Also examine the percentage of each of the primal terms in the objective function and in the original objective function.

2. The constraint is a binding constraint. If the constraint is removed, the objective function should be lower. What problem(s) occurs when the constraint is removed that causes concern?

11.3 REFERENCES

[1] `http://www.mpri.lsu/edu/textbook/Chapter3-b.htm` (visited May 2009). 63

[2] T. K. Sherwood, *A Course in Process Design*, MIT Press, Cambridge, MA, 1963. 63

CHAPTER 12

Material Removal/Metal Cutting Economics Case Study

12.1 INTRODUCTION

Material removal economics, also known as metal cutting economics or machining economics, is an example of a problem which has non-integer exponents that can be varied, and this makes the problem challenging. This problem has been presented previously [1, 2], but this version is slightly different from and easier than those presented earlier. The material removal economics problem is based upon the Taylor Tool Life Equation which was developed by Frederick W. Taylor over 100 years ago in the U.S. There are several versions of the equation, and the form selected is one of the modified versions which includes cutting speed and feed rate. The equation selected was:

$$TV^{1/n}f^{1/m} = C, \tag{12.1}$$

where

T = tool life (minutes)

V = cutting speed (ft/min or m/min)

F = feed rate (inches/rev or mm/rev)

$1/n$ = cutting speed exponent

$1/m$ = feed rate exponent

C = Taylor's Modified Tool Life Constant (ft/min or m/min).

The object is to minimize the total cost for cutting cost, loading and unloading cost, tool cost, and tool changing cost per unit.

12.2 PROBLEM FORMULATION

An expression for the loading and unloading cost, cutting cost, tool changing cost, and tool cost was developed [1] and the resulting expression was:

$$C_u = \text{loading and unloading cost} + \text{cutting cost}$$
$$+ (\text{tool changing cost} + \text{tool cost}) \tag{12.2}$$
$$C_u = K_{00} + K_{01}f^{-1}V^{-1} + K_{02}f^{(1/m-1)}V^{(1/n-1)}, \tag{12.3}$$

where

$$C_u = \text{total unit cost}$$
$$K_{00} = (R_o + R_m)t_l$$
$$K_{01} = (R_o + R_m)B$$
$$K_{02} = [(R_o + R_m)t_{ch} + C_t]QBC^{-1},$$

and

R_o = operator rate (\$/min), derived from Operator's Hourly Cost

R_m = machine rate (\$/min), derived from Machine Hour Cost

t_l = machine loading and unloading time (min)

t_{ch} = tool changing time (min)

B = cutting path surface factor of tool (in-ft, or mm-m)

Q = fraction of cutting path that tool is cutting material

C_t = tool cost (\$/cutting edge), derived from Tool Insert Cost

C = Taylor's Modified Tool Life Constant (min).

The first part of the objective function expression represents the loading and unloading costs, the second part represents the cutting costs, and the third part represents the tool and tool changing costs. The loading and unloading costs are not a function of the feed and cutting speed. Since K_{00} is a constant, the primal problem can be formulated as solving for the variable cost, $C_u(var)$ as:

$$C_u(var) = K_{01}f^{-1}V^{-1} + K_{02}f^{(1/m-1)}V^{(1/n-1)}. \tag{12.4}$$

Subject to a maximum feed constraint written as:

$$K_{11}f \leq 1, \tag{12.5}$$

where

$$K_{11} = 1/f_{\max}.$$

The degrees of difficulty have been reduced to zero with the elimination of the constant term as the number of terms is reduced from 4 to 3 and the degrees of difficulty become:

$$D = T - (N + 1) = 3 - (2 + 1) = 0. \tag{12.6}$$

From the coefficients and signs, the signum values for the dual are:

$$\sigma_{01} = 1$$
$$\sigma_{02} = 1$$
$$\sigma_{11} = 1$$
$$\sigma_1 = 1.$$

The dual problem formulation is:

$$
\begin{array}{llll}
\text{Objective Function} & \omega_{01} & + \omega_{02} & = 1 & (12.7) \\
f \text{ terms} & -\omega_{01} + (1/m - 1)\omega_{02} + \omega_{11} & = 0 & (12.8) \\
V \text{ terms} & -\omega_{01} + (1/n - 1)\omega_{02} & = 0. & (12.9)
\end{array}
$$

From the constraint equations which have only one term it is apparent that:

$$\omega_{10} = \omega_{11}. \tag{12.10}$$

Since there are zero degrees of difficulty, the dual parameters can be solved for directly. Thus, if one adds Equations (12.7) and (12.9) one can solve for ω_{02} directly and obtain:

$$\omega_{02} = n. \tag{12.11}$$

Then from Equation (12.7) one obtains

$$\omega_{01} = 1 - n. \tag{12.12}$$

Finally, by using the values for ω_{01} and ω_{02} one can determine ω_{11} as:

$$\omega_{11} = 1 - n/m. \tag{12.13}$$

The objective function can be evaluated using the dual expression:

$$C_u(var) = d(\omega) = \sigma \left[\prod_{m=0}^{M} \prod_{t=1}^{T_m} (C_{mt}\omega_{mo}/\omega_{mt})^{\sigma_{mt}\omega_{mt}} \right]^{\sigma} \tag{12.14}$$

$$
\begin{aligned}
C_u(var) &= 1\left\{[(K_{01}\omega_{00}/\omega_{01})\omega_{01}]\left[(K_{02}\omega_{00}/\omega_{02})\omega_{02}\right]\left[(K_{11}\omega_{10}/\omega_{11})\omega_{11}\right]\right\}^1 \tag{12.15}\\
&= \left[(K_{01}1/(1-n))^{(1-n)}\right]\left[(K_{02}1/n)^n\right]\\
&\quad \left[(K_{11}(1-n/m)/(1-n/m)^{(1-n/m)}\right]\\
&= [K_{01}/(1-n)][(K_{02}/K_{01})((1-n)/n)]^n \left[(K_{11})^{(1-n/m)}\right]\\
&= (K_{11})^{1-n/m} K_{01}^{1-n} K_{02}^{n}(1-n)^{n-1}/n^n. \tag{12.16}
\end{aligned}
$$

The primal variables, V and f, can be evaluated from the primal-dual relationships.

$$K_{01}f^{-1}V^{-1} = \omega_{01}C_u(var) \tag{12.17}$$

$$K_{02}f^{1/m-1}V^{1/n-1} = \omega_{02}C_u(var) \tag{12.18}$$

$$K_{11}f = 1. \tag{12.19}$$

From Equation (12.19), it is seen that:

$$f = 1/K_{11}. \tag{12.20}$$

If one divides Equation (12.18) by Equation (12.17), one obtains:

$$f^{1/n} V^{1/m} = (K_{01}/K_{02})(n/(1-n)). \tag{12.21}$$

Using Equation (12.20) in (12.21) and solving for V one obtains

$$V = [(n/(1-n))]^n (K_{01}/K_{02})^n K_{11}^{n/m}. \tag{12.22}$$

Now if one uses the values of f and V from Equations (12.20) and (12.22) in the primal Equation (12.4) and also using Equation (12.21) one obtains:

$$\begin{aligned}
C_u(var) &= K_{01} f^{-1} V^{-1} + K_{02} f^{(1/m-1)} V^{(1/n-1)} \tag{12.4} \\
&= f^{-1} V^{-1} \left[K_{01} + K_{02} f^{1/m} V^{1/n} \right] \\
&= K_{11}[(n/(1-n))^{-n} (K_{01}/K_{02})^{-n} K_{11}^{-n/m}[K_{01} + K_{02}(n/(1-n))(K_{01}/K_{02})] \\
&= K_{11}^{1-n/m}(n/(1-n))^{-n}(K_{01}/K_{02})^{-n}[K_{01} + K_{01}(n/(1-n))] \\
&= K_{11}^{1-n/m} K_{01}^{1-n} K_{02}^{n}(1-n)^{n-1}/n^n. \tag{12.23}
\end{aligned}$$

The variable unit cost expressions, $C_u(var)$, are identical for both the primal and dual formulations. The expressions for the primal variables and the variable unit cost are more complex than the expressions obtained in the previous models because of the non-integer exponents. One must also note that the dual variables are a function of the exponents n and m, which can change for different materials can cutting tools.

Elaborate research work has been done with the material removal problems, and the dissertation by Pingfang Tsai [3] has solutions for problems with an additional variable, the depth of cut, and additional constraints on the horsepower of the motor driving the main spindle of the lathe and depth of cut. With the additional constraints, there is the possibility of loose constraints, and a flow chart has been developed for the different solutions depending upon which constraints are loose. Chapter 21 presents a material removal problem with one degree of difficulty and two constraints.

12.3   EVALUATIVE QUESTIONS

1. A cylindrical bar, 6 inches long and 1 inch in diameter, is to be finished turned on a lathe. The maximum feed to be used to control the surface finish is 0.005 in/rev. Find the total cost to machine the part, the variable cost to machine the part, the feed rate, the cutting speed, and the tool life in minutes. Use both the primal and dual equations to determine

the variable unit cost. The data are:

$$R_o = 0.60 \text{ \$/min}$$
$$R_m = 0.40 \text{ \$/min}$$
$$C_t = \text{\$2.00/edge}$$
$$t_l = 1.5 \text{ min}$$
$$t_{ch} = 0.80 \text{ min}$$
$$D = 1 \text{ inch}$$
$$L = 6 \text{ inches}$$
$$B = \pi DL = \pi * 1 \text{ in} * 6 \text{ in/rev} * 1 \text{ ft/12 in} = \pi/2 \text{ (in-ft/rev)} = 1.57 \text{ (in-ft/rev)}$$
$$1/m = 1.25 \ (m = 0.80)$$
$$1/n = 4.00 \ (n = 0.25)$$
$$C = 5.0 \times 10^8 \text{ min}$$
$$Q = 1.0 \text{ (for turning)}.$$

Using these values one can obtain:

$$K_{00} = (R_o + R_m)t_l = (0.60 \text{ \$/min} + 0.40 \text{ \$/min}) 1.50 \text{ min} = 1.50 \text{ \$}$$
$$K_{01} = (R_o + R_m)B = (0.60 \text{ \$/min} + 0.40 \text{ \$/min}) 1.57 \text{ in-ft/rev}$$
$$= 1.57 \text{ (\$-in-ft/(min-rev))}$$
$$K_{02} = [(R_o + R_m)t_{ch} + C_t]QBC^{-1}$$
$$= [(0.60 \text{ \$/min} + 0.40 \text{ \$/min}) 0.80 \text{ min} + 2.00 \text{ \$}]$$
$$(1.0)(1.57 \text{ in-ft/rev})/(5.0 \times 10^8 \text{ min})$$
$$= 8.8 \times 10^{-9} \text{ ((\$-in-ft)/(min-rev))}$$

(solution $f = 0.005$ in/rev, $V = 459$ ft/min, $C_u(var) = 0.91$, and $T = 8.5$ min).

2. A cylindrical bar, 150 mm long and 25 mm in diameter, is to be finished turned on a lathe. The maximum feed to be used to control the surface finish is 0.125 mm/rev. Find the total cost to machine the part, the variable cost to machine the part, the feed rate, the cutting

speed, and the tool life in minutes. The data are:

$$R_o = 0.60 \ \$/\text{min}$$
$$R_m = 0.40 \ \$/\text{min}$$
$$C_t = \$2.00/\text{edge}$$
$$t_l = 1.5 \ \text{min}$$
$$t_{ch} = 0.8 \ \text{min}$$
$$D = 25 \ \text{mm}$$
$$L = 150 \ \text{mm}$$
$$B = \pi DL = \pi * 25 \ \text{mm/rev} * 150 \ \text{mm} * 1 \ \text{m}/1000 \ \text{mm}$$
$$= 3.75\pi \ (\text{m-mm/rev}) = 11.78 \ (\text{m-mm/rev})$$
$$1/m = 1.25 \ (m = 0.80)$$
$$1/n = 4.00 \ (n = 0.25)$$
$$C = 2.46 \times 10^8 \ \text{min}$$
$$Q = 1.0 \ (\text{for turning}).$$

Using these values one can obtain:

$$K_{00} = (R_o + R_m)t_l = (0.60 \ \$/\text{min} + 0.40 \ \$/\text{min}) \ 1.50 \ \text{min} = 1.50 \ \$$$
$$K_{01} = (R_o + R_m)B = (0.60 \ \$/\text{min} + 0.40 \ \$/\text{min}) \ 11.78 \ (\text{m-mm/rev})$$
$$= 11.78 \ (\$\text{-m-mm}/(\text{min-rev}))$$
$$K_{02} = [(R_o + R_m)t_{ch} + C_t]QBC^{-1}$$
$$= [(0.60 \ \$/\text{min} + 0.40 \ \$/\text{min}) \ 0.80 \ \text{min} + 2.00]$$
$$(1.0)(11.78 \ \text{m-mm/rev})/(2.46 \times 10^8 \ \text{min})$$
$$= 1.34 \times 10^{-7}((\$\text{-m-mm})(\text{min-rev}))$$

(solution $f = 0.125$ mm/rev, $V = 140$ m/min, $C_u(var) = 0.90$, and $T = 8.6$ min).

12.4   REFERENCES

[1] R. C. Creese and P. Tsai, Generalized solution for constrained metal cutting economics problem, *Annual International Industrial Conference Proceedings*, Institute of Industrial Engineers U.S., pp. 113–117, 1985. 69

[2] D. S. Ermer, Optimization of the constrained machining economics problem by geometric programming, *Journal of Engineering for Industry*, Transactions of the ASME, pp. 1067–1072, 1971. 69

[3] P. Tsai, *An Optimization Algorithm and Economic Analysis for a Constrained Machining Model*, p. 214, Ph.D. Dissertation, West Virginia University, Morgantown, WV. 72

CHAPTER 13

Construction Building Sector Cost Minimization Case Study

13.1 INTRODUCTION

Production functions are frequently used in the area of economic models. A model for the civil engineering building construction sector in Turkey was developed by Guney and Oz [1]. The model utilizes a Cobb-Douglas production function and minimizes the cost of the inputs used to produce a specific production level in the building construction sector.

The outputs of the model are the minimum amounts of capital and of labor to produce a given level of production for specific capital and labor rates. A slightly modified version of the model is presented [2] which gives the same results as the original model of Guney and Oz and develops estimating equations for the labor amount required and the capital amount required.

13.2 MODEL DEVELOPMENT

The objective function for the model is to minimize the total costs subject to the Cobb-Douglas production constraint. The objective function is:

$$Y(x) = r_1 x_1 + r_2 x_2 \tag{13.1}$$

subject to the Cobb-Douglas production constraint

$$q = A x_1^\alpha x_2^\beta, \tag{13.2}$$

where

x_1 = labor amount

x_2 = capital amount

r_1 = labor rate

r_2 = capital rate

q = desired output level

A = total productivity factor

α = labor elasticity

β = capital elasticity.

The constraint must be written in the inequality form and becomes:

$$(q/A)x_1^{-\alpha}x_2^{-\beta} \le 1. \tag{13.3}$$

Thus, the primal problem is given by the objective function of Equation (13.1) and the constraint by Equation (13.3).

The degrees of freedom are found to be:

$$D = T - (N + 1) = 3 - (2 + 1) = 0 \tag{13.4}$$

where

$T =$ Number of terms of primal

$N =$ number of orthogonal conditions (also the number of variables in primal).

The dual formulation initially appears more complex, but it results in linear equations which are easier to solve. The dual objective function is not linear and is solved after the dual variables have been determined from the dual formulation model. The dual objective function is:

$$d(\omega) = \sigma \left[\prod_{m=0}^{M} \prod_{t=1}^{T_m} (C_{mt}\omega_{m0}/\omega_{mt})^{\sigma_{mt}\omega_{mt}} \right]^{\sigma} \tag{13.5}$$

for

$$m = 0, 1, 2, \ldots, M \quad \text{and} \quad t = 1, 2, \ldots, T_m,$$

where

$\sigma =$ signum function for objective function (1 for minimization)

$\sigma_{mt} =$ signum function for dual constraints (± 1)

$C_{mt} > 0$ positive constant coefficients

$\omega_{m0} =$ dual variables from the linear inequality constraints

$\omega_{mt} =$ dual variables of dual constraints

$\sigma_{mt} =$ signum function for dual constraints

$\omega_{00} = 1.$

Since all the terms in Equations (13.1) and (13.3) are positive, the signum values are all positive, that is

$\sigma_{00} = 1$ (objective function is minimization)

$\sigma_{01} = 1$

$\sigma_{02} = 1$

$\sigma_{11} = 1$

$\sigma_{10} = 1$ (RHS of constraint is positive).

The dual can be formulated as:

$$\text{(primal objective function terms)} \quad \omega_{01} + \omega_{02} \qquad = 1 \tag{13.6}$$

$$\text{(primal variable } x_1 \text{ terms)} \quad \omega_{01} \qquad -\alpha\omega_{11} = 0 \tag{13.7}$$

$$\text{(primal variable } x_2 \text{ terms)} \quad \omega_{01} \qquad -\beta\omega_{11} = 0. \tag{13.8}$$

Solving Equations (13.6)–(13.8) for the dual variables one obtains:

$$\omega_{01} = (\alpha/(\alpha + \beta)) \tag{13.9}$$

$$\omega_{02} = (\beta/(\alpha + \beta)) \tag{13.10}$$

$$\omega_{11} = (1/(\alpha + \beta)). \tag{13.11}$$

Now ω_{10} can be determined using

$$\omega_{10} = \sigma_{10} \sum \sigma_{mt}\omega_{mt} \tag{13.12}$$

$$= 1 * (1 * (1/(\alpha + \beta)))$$

$$= 1/(\alpha + \beta). \tag{13.13}$$

The dual objective function of Equation (13.5) can now be determined and is:

$$d(\omega) = 1 * \left[\{r_1 * 1/(\alpha/(\alpha + \beta))\}^{(1*(\alpha/(\alpha+\beta)))} \right.$$

$$\left. * \{r_2 * 1/(\beta/(\alpha + \beta))\}^{(1*(\beta/(\alpha+\beta)))} * \{q/A\}^{(1*(1/(\alpha+\beta)))} \right]^{1}. \tag{13.14}$$

Now that the dual variables are determined, the primal variables can be determined from the relationships between the primal and dual variables. As in linear programming, the primal and dual objective functions must be equal and thus $Y_0(x)$ and $d(\omega)$ are equal. The two equations relating the primal and dual for determining the primal variables are:

$$C_{0t} \prod_{n=1}^{N} x_n^{mtn} = \omega_{0t}\sigma d(\omega) \tag{13.15}$$

and

$$C_{mt} \prod_{n=1}^{N} x_n^{mtn} = \omega_{mt}/\omega_{m0} \tag{13.16}$$

$$\text{for} \quad t = 1, 2, \ldots, T_m \quad \text{and} \quad m = 1, 2, \ldots, M.$$

Remember that both ω_{11} and ω_{10} are equal. Now using the primal-dual relationship of Equation (13.15) for the two terms of the objective function, one obtains:

$$r_1 x_1 = (\alpha(\alpha + \beta)) * 1 * d(\omega) \tag{13.17}$$

$$r_2 x_2 = (\beta(\alpha + \beta)) * 1 * d(\omega). \tag{13.18}$$

Solving Equations (13.17) and (13.18) for x_1 one obtains:

$$x_1 = (\alpha/\beta)(r_2/r_1)x_2. \tag{13.19}$$

Now using Equation (13.17) in Equation (13.2) and solving for x_2:

$$x_2 = (q/A)^{(1/(\alpha+\beta))}(\alpha r_2/\beta r_1)^{(-\alpha/(\alpha+\beta))}. \tag{13.20}$$

Using Equation (13.20) in Equation (13.19) x_1 is found to be

$$x_1 = (q/A)^{(1/(\alpha+\beta))}(\alpha r_2/\beta r_1)^{(\beta(\alpha+\beta))}. \tag{13.21}$$

The primal objective function can now be determined from the primal variables and Equation (13.1) becomes

$$Y(x) = r_1 * \left[(q/A)^{(1/(\alpha+\beta))}(\alpha r_2/\beta r_1)^{(\beta(\alpha+\beta))}\right]$$
$$+ r_2 * \left[(q/A)^{(1/(\alpha+\beta))}(\alpha r_2/\beta r_1)^{(-\alpha/(\alpha+\beta))}\right]. \tag{13.22}$$

Equations (13.14) and (13.22) have quite different appearances, but the numerical values will be the same. Equations (13.20) and (13.21) are considered to be design equations as they are in terms of the input constants and the problem does not need to be resolved, but only the design equations need to be evaluated with the changed constants. This permits it to be relatively easy to evaluate sensitivity of the design parameter to the changes in the constants.

13.3   MODEL RESULTS AND VALIDATION

The equations developed were used to compare with the data reported by Guney and Oz [1] on the construction sector in Turkey. The input values are given in Table 13.1 and the output values are in Table 13.2. The input values of A, α, and β were fixed at 1.0, 0.53, and 0.47 for all four reported cases.

The results for x_1 and x_2, the labor and capital estimates, were in complete agreement with those of Guney and Oz [1]. The complete set of results are presented in Table 13.2. The primal and dual values of the objective function from Equations (13.14) and (13.22) are identical as expected and although the labor cost, capital cost, and objective function were not given in the reference [1], they would most likely have been the same. In the model equations developed, the values for A, α, and β from the Cobb-Douglas production equation could be varied, and a

Table 13.1: Yearly input data for estimate calculations

Year	Production Index q	Labor Index r_1	Capital index r_2
2006	118.4	121.88	114.32
2007	124.9	137.80	122.32
2008	115.6	153.85	140.06
2009	96.4	158.53	131.48

Table 13.2: Output values of model for estimates

Year	Output Level q	Labor Estimate x_1	Capital Estimate x_2	Labor Rate r_1	Capital Rate r_2	Labor Cost $r_1 x_1$	Capital Cost $r_2 x_2$	Primal Total Cost (Y)	Dual Total Cost d(ω)
2006	118.4	121.56	114.93	121.88	114.32	14,816	13,139	27,955	27,955
2007	124.9	124.96	124.84	137.80	122.32	17,219	15,270	32,489	32,489
2008	115.6	117.03	114.00	153.85	140.06	18,006	15,967	33,973	33,973
2009	96.4	93.41	99.88	158.53	131.48	14,809	13,132	27,941	27,941

sensitivity analysis of these parameters could be evaluated and would not require resolving the adjusted problem.

Note that the labor cost and capital cost ratios to the total cost can be determined from the dual variables in Equations (13.9) and (13.10). The labor cost ratio is:

$$\omega_{01} = (\alpha/(\alpha + \beta)) = (0.53/(0.53 + 0.47)) = 0.53. \tag{13.9}$$

The capital cost ratio is:

$$\omega_{02} = (\beta/(\alpha + \beta)) = (0.47/(0.53 + 0.47)) = 0.47. \tag{13.10}$$

The dual variables are ratios of the two elastic parameters. This implies that for the specific values of α and β given, the labor cost will be 53% of the total cost and the capital cost will be 47% of the total cost and only the amounts of the costs will be effected by changes in r_1, r_2, q, and A.

13.4 CONCLUSIONS

The development of design equations for a geometric programming model of data from the construction sector in Turkey. These design equations [Equations (13.20) and (13.21)] give the solutions for the model outputs and the model does not need to be resolved. The design equations for the variables x_1 and x_2 can then be used to determine the total cost to meet the desired production level. The design equations also permit easy analysis of the impact of the Cobb-Douglas elasticity exponents and total productivity factor upon the total cost. The development of design equations takes considerable effort, but the equations permit a more rapid analysis of the impact of the input variables upon the output.

13.5　EVALUATIVE QUESTIONS

1.　a. Resolve the example problem with the new values of $\alpha = 0.55$, $\beta = 0.48$, and $A = 1.2$ using the equations developed for x_1 and x_2 and the initial values of q, r_1, and r_2.

　　b. Show that two terms of the primal have the same ratio to the total cost as the values of the corresponding dual variables ω_{01} and ω_{02}.

2. If the elasticity values are $\alpha = 0.58$ and $\beta = 0.48$, what is the expected ratio of the total labor costs to the total capital costs? What are the total cost, the labor cost, and the capital cost values?

3. Show that Equations (13.14) and (13.22) are equivalent. (A challenge question.)

13.6　REFERENCES

[1] I. Guney, and E. Oz, An application of geometric programming, *International Journal of Electronics, Mechanical, and Mechatronics Engineering*, Vol. 2, pp. 157–161, 2012. 77, 80

[2] R. C. Creese, Design equations from geometric programming, *International Journal of Electronics, Mechanical, and Mechatronics Engineering*, Vol. 5, no. 2, pp. 963–968. 77

PART III

Geometric Programming Profit Maximization Applications with Zero Degrees of Difficulty

CHAPTER 14

Production Function Profit Maximization Case Study

14.1 PROFIT MAXIMIZATION WITH GEOMETRIC PROGRAMMING

The examples thus far have been minimization problems, and the maximization problems are slightly different. In minimization problems, the terms of the objective function have only positive signs or all positive signum functions. The constraints may have negative signum functions in a minimization problem, but those in the objective function are all positive. In the profit maximization problem, the revenues have positive coefficients and the costs have negative coefficients. The solution formulation for the maximization problem is slightly different than the minimization problem, and a solution to an example is presented to illustrate the problem formulation and solution procedure.

The formulation of the dual objective function for profit maximization is similar to that of cost minimization, but the sign of the sigmum fuctions are negative as illustrated in Equation (14.1).

The dual objective function is expressed as:

$$d(\omega) = \sigma \left[\prod_{m=0}^{M} \prod_{t=1}^{T_m} (C_{mt}\omega_{m0}/\omega_{mt})^{\sigma_{mt}\omega_{mt}} \right]^{\sigma}$$

$$m = 0, 1, \ldots, M \text{ and } t = 1, 2, \ldots, T_m, \tag{14.1}$$

where

σ = signum function (-1 for maximization and $+1$ for minimization)

C_{mt} = constant coefficient

ω_{m0} = dual variables from the linear inequality constraints

ω_{mt} = dual variables of dual constraints, and

σ_{mt} = signum function for dual constraints.

Thus, the two major differences in the formulation of a maximization problem from that of a minimization problem are:

1. Sign changes in the objective function will occur. Thus, the sum of the dual variables of the objective function will be greater than unity and will not represent a simple sum of the cost terms. However, the cost ratios can still be determined.

2. The value of σ in the formulation will be negative one (-1) instead of the positive one $(+1)$ used for minimization.

These differences will be emphasized in the production function profit maximization problem.

14.2 PROFIT MAXIMIZATION OF THE PRODUCTION FUNCTION CASE STUDY

This is a modification of a problem used on an old exam. The production function is an expression of capital (C) and labor (L). The specific production function is given by $C^{0.3}L^{0.4}$ production units and the revenue per unit is \$100. The capital cost per unit is \$12 and the labor cost per unit is \$10.

a. What is the total profit per unit?

b. What is the total revenue?

c. What is the total capital cost?

d. What is the total labor cost?

The primal objective function expression can be expressed as:

$$\text{Profit}(\pi) = \text{Revenue} - \text{capital costs} - \text{labor cost}$$
$$\pi = 100C^{0.3}L^{0.5} - 12C - 10L. \tag{14.2}$$

For profit maximization, the primal objective function is expressed as:

$$-\pi = -100C^{0.3}L^{0.5} + 12C + 10L, \tag{14.3}$$

where

C = capital units

L = labor units

α = capital elasticity factor = 0.3

β = labor elasticity factor = 0.5.

The values of the capital elasticity factor are generally between 0.40 and 0.85 and the values of the labor elasticity factor are generally between 0.15 and 0.60 according to Nijkamp [1].

The degrees of freedom are obtained by

$$D = T - (N + 1) = 3 - (2 + 1) = 0, \tag{14.4}$$

where

T = number of terms in primal

N = number of variables in primal.

The dual expression from Equation (14.1) is repeated as:

$$d(\omega) = \sigma \left[\prod_{m=0}^{M} \prod_{t=1}^{T_m} (C_{mt}\omega_{m0}/\omega_{mt})^{\sigma_{mt}\omega_{mt}} \right]^{\sigma}$$
$$m = 0, 1, \ldots, M \quad \text{and} \quad t = 1, 2, \ldots T_m \tag{14.1}$$

where

σ = signum function for objective function (-1 for maximization)

σ_{mt} = signum function for dual constraints (± 1)

$C_{mt} > 0$ positive constant coefficients

ω_{m0} = dual variables from the linear inequality constraints

ω_{mt} = dual variables of dual constraints

σ_{mt} = signum function for dual constraints

$\omega_{00} = 1.$

Since not all of the terms in Equations (14.3) have positive signs, the signum values are not all positive, that is

$$\sigma_{00} = -1 \text{ (objective function is minimization)}$$
$$\sigma_{01} = -1$$
$$\sigma_{02} = 1$$
$$\sigma_{03} = 1.$$

The dual formulation would be:

(primal objective function terms)	$-\omega_{01} + \omega_{02} + \omega_{03} = -1$	(14.5)
(primal variable C terms)	$-0.3\omega_{01} + \omega_{02} = 0$	(14.6)
(primal variable L terms)	$-0.5\omega_{01} + \omega_{03} = 0.$	(14.7)

Solving Equations (14.5)–(14.7) for the dual variables, one obtains

$$\omega_{01} = 1/[1 - (\alpha + \beta)] = 1/[1 - (0.3 + 0.5)] = 1/0.2 = 5 \qquad (14.8)$$
$$\omega_{02} = \alpha/[1 - (\alpha + \beta)] = 0.3/[1 - (0.3 + 0.5)] = 0.3/0.2 = 1.5 \qquad (14.9)$$
$$\omega_{03} = \beta/[1 - (\alpha + \beta)] = 0.5/[1 - (0.3 + 0.5)] = 0.5/0.2 = 2.5. \qquad (14.10)$$

The dual objective function would be:

$$d(\omega) = -1\left[(C_{01}/\omega_{01})^{-\omega_{01}}(C_{02}/\omega_{02})^{\omega_{02}}(C_{03}/\omega_{03})^{\omega_{03}}\right]^{-1} \qquad (14.11)$$
$$= -1\left[(100/5)^{-5}(12/1.5)^{1.5}(10/2.5)^{2.5}\right]^{-1} = -1\left[2.2627 \times 10^{-4}\right]^{-1}$$
$$= -4{,}419.42.$$

The negative sign implies that it is a profit of \$4,419 rather than as a negative cost. Now that the dual variables are determined, the primal variables can be determined from the relationships between the primal and dual variables. Note that the ratio between the two cost terms, ω_{02} and ω_{03}, are in the ratio of 0.3:0.5. That is, the capital costs will be 60% of the labor costs and the total revenue will be twice the labor costs, and the profit will be equal to the difference between the labor and capital costs for this problem. The equations relating the primal and dual for determining the primal variables are:

$$C_{0t} \prod_{n=1}^{N} x_n^{mtn} = \omega_{0t}\sigma d(\omega). \qquad (14.12)$$

Using the third term

$$10L = (2.5)(-1)(-4{,}419.42) = 11{,}048.5$$
$$L = 11{,}048.5/10 = 1{,}104.85. \qquad (14.13)$$

Using the second term

$$12C = (1.5)(-1)(-4{,}419.42) = 6{,}629.13$$
$$C = 6{,}629.13/12 = 552.43. \qquad (14.14)$$

Using Equation (14.2) to calculate the profit from the primal expression, one obtains:

$$\pi = 100C^{0.3}L^{0.5} - 12C - 10L \qquad (14.2)$$
$$= 100(552.43)^{0.3}(1104.85)^{0.5} - 12 * (552.43) - 10 * (1104.85)$$
$$= 22{,}097.07 - 6{,}629.16 - 11{,}048.5$$
$$\pi = 4{,}419.42. \qquad (14.15)$$

Thus, the primal and dual objective functions are in agreement. If one examines the dual variables as a ratio, note that for this problem they represent the ratio of the

Revenues : Capital Costs : Labor Costs : Profits and $\omega_{01} : \omega_{02} : \omega_{03} : \pi$

22,097.07 :   6,629.16 :   11,048.5   :4,419.4 and   5   : 1.5 : 2.5 : 1.0.

Note that the values of the dual variables for the cost terms, ω_{02} and ω_{03}, do not sum to unity as they did in the cost minimization problems. However, the difference between the revenues and the sum of the cost terms, which is the profits, does equal unity. The profit maximization case is slightly more difficult than the cost minimization case, but the importance of the dual and its cost proportions is still maintained.

Since the dual variables do not change with the coefficients, design equations can be developed in terms of the cost coefficients. With the dual variables of 5, 1.5, and 2.5 the dual objective function can be represented as a profit by:

$$d(\omega) = 1\left[(C_{01}/5)^{-5}(C_{02}/1.5)^{1.5}(C_{03}/2.5)^{2.5}\right]^{-1}. \tag{14.16}$$

Using the primal-dual relationships, the equations for C and L can be obtained as:

$$C = (1.5/C_{02})d(\omega) \tag{14.17}$$
$$L = (2.5/C_{03})d(\omega). \tag{14.18}$$

The values from the current problem in Equations (14.16)–(14.18) result in:

$$\begin{aligned}
d(\omega) &= 1\left[(C_{01}/5)^{-5}(C_{02}/1.5)^{1.5}(C_{03}/2.5)^{2.5}\right]^{-1} \\
&= 1\left[(100/5)^{-5}(12/1.5)^{1.5}(10/2.5)^{2.5}\right]^{-1} \\
&= 4419 \\
C &= (1.5/C_{02})d(\omega) \\
&= (1.5/12)4419 \\
&= 552.4 \\
L &= (2.5/C_{03})d(\omega) \\
&= (2.5/10)4419 \\
&= 1104.9.
\end{aligned}$$

Since the dual variables are constant as long as the equation exponents do not change, the design equations can be further simplified to:

$$d(\omega) = (5.8095 * 10^{-3}) * \left[(C_{01})^5 / ((C_{02})^{1.5} * (C_{03}^{2.5}))\right] \tag{14.19}$$
$$C = (8.7142 * 10^{-3}) * \left[(C_{01})^5 / ((C_{02})^{2.5} * (C_{03}^{2.5}))\right] \tag{14.20}$$
$$L = (1.4524 * 10^{-2}) * \left[(C_{01})^5 / ((C_{02})^{1.5} * (C_{03}^{3.5}))\right]. \tag{14.21}$$

The design equations permit a relatively easy method to determine the effect of the cost coefficients upon the objective function and the design variables and one does not need to resolve the original problem unless the equation exponents change. However, if the exponents change, the dual variables would change and new design equations would need to be developed.

14.3 EVALUATIVE QUESTIONS

1. Change the coefficients of C_{01}, C_{02}, and C_{03} to 50, 18, and 20 and determine the amount of profit and revenue and the values of C and L. Also show that they follow the ratios of the dual variables.

2. Change the exponent of C from 0.3 to 0.4 and the values of C_{01}, C_{02}, and C_{03} to 40, 16, and 5. Solve for the amount of profit, revenue, and the values of C and L.

3. What happens to the dual variables if $\alpha = 0.6$ and $\beta = 0.5$?

14.4 REFERENCES

[1] P. Nijkamp, *Planning of Industrial Complexes by Means of Geometric Programming*, p. 11, Rotterdam University Press, 1972. 86

CHAPTER 15

Product Mix Profit Maximization Case Study

15.1 PROFIT MAXIMIZATION USING THE COBB-DOUGLAS PRODUCTION FUNCTION

The problem selected was presented by Liu [1] who adapted it from Keyzer and Wesenbeeck [2]. The objective is to determine the product mix which maximizes the profit, π, using a Cobb-Douglas production function and the component costs. The problem given is

$$\text{Maximize } \pi = pAx_1^{0.1}x_2^{0.3}x_3^{0.2} - C_{01}x_1 - C_{02}x_2 - C_{03}x_3, \tag{15.1}$$

where the parameters are in Table 15.1.

Table 15.1: Model parameters for product mix profit maximization problem

Parameter	Term	Value for Problem
ρ	= market price	= 20
A	= scale of production (Cobb-Douglas function)	= 40
C_{01}	= Cost of Product 1	= 20
C_{02}	= Cost of Product 2	= 24
C_{03}	= Cost of Product3	= 30
α	= cost elasticity of product 1	= 0.1
β	= cost elasticity of product 2	= 0.3
γ	= cost elasticity of product 3	= 0.2
C_{00}	= ρ A	= 800
x_1	= Amount of product 1	
x_2	= Amount of product 2	
x_3	= Amount of product 3	
π	= Amount of Profit	

To solve the problem, one minimizes the negative of the profit function; that is the primal objective function is:

$$\text{Minimize } Y = -C_{00}x_1^{0.1}x_2^{0.3}x_3^{0.2} + C_{01}x_1 + C_{02}x_2 + C_{03}x_3, \tag{15.2}$$

were

$$Y = -\pi.$$

The degrees of freedom are obtained by

$$D = T - (N + 1) = 4 - (3 + 1) = 0, \tag{15.3}$$

where

T = number of terms in primal

N = number of variables in primal.

The dual objective function would be:

$$D(Y) = -1\left[(C_{00}/\omega_{01})^{-\omega_{01}}(C_{01}/\omega_{02})^{\omega_{02}}(C_{02}/\omega_{03})^{\omega_{03}}(C_{03}/\omega_{04})^{\omega_{04}}\right]^{-1}. \tag{15.4}$$

For maximization of profits, the signum function $\sigma_{00} = -1$, whereas in the dual objective function for the minimization of costs the signum function $\sigma_{00} = 1$, a positive value. From the coefficients and signs, the signum values for the dual are:

$$\sigma_{01} = -1$$
$$\sigma_{02} = 1$$
$$\sigma_{03} = 1$$
$$\sigma_{04} = 1.$$

The dual formulation would be:

Objective Fctn	$-\omega_{01} + \omega_{02} + \omega_{03} + \omega_{04} = -1$ (max)	(15.5)
x_1 terms	$-\alpha\omega_{01} + \omega_{02} \phantom{+\omega_{03} + \omega_{04}} = 0$	(15.6)
x_2 terms	$-\beta\omega_{01} \phantom{+\omega_{02}} + \omega_{03} \phantom{+ \omega_{04}} = 0$	(15.7)
x_3 terms	$-\gamma\omega_{01} \phantom{+\omega_{02} +\omega_{03}} + \omega_{04} = 0.$	(15.8)

Solving Equations (15.5) through (15.8) for the dual variables one obtains:

$$\omega_{01} = 1/[1 - (\alpha + \beta + \gamma)] = 1/[1 - (0.1 + 0.3 + 0.2)] = 1/0.4 = 2.50 \tag{15.9}$$
$$\omega_{02} = \alpha\omega_{01} = 0.1 * 2.5 = 0.25 \tag{15.10}$$
$$\omega_{03} = \beta\omega_{01} = 0.3 * 2.5 = 0.75 \tag{15.11}$$
$$\omega_{04} = \gamma\omega_{01} = 0.2 * 2.5 = 0.50. \tag{15.12}$$

The dual variables do not sum to unity which was the case for the cost minimization problems. However, the difference between the profit dual variable and cost dual variables is equal to unity. The dual variables indicate that there will always be a profit, that the revenue will be 5/3 the sum of the costs and that the profit will be 2/3 that of the total costs. Also, the cost terms should be in the same ratio as their corresponding dual variables. The dual objective function $D(Y)$ would result:

$$D(Y) = -1\left[(C_{00}/\omega_{01})^{-\omega_{01}}(C_{01}/\omega_{02})^{\omega_{02}}(C_{02}/\omega_{03})^{\omega_{03}}(C_{03}/\omega_{04})^{\omega_{04}}\right]^{-1}$$

$$= -1\left[(C_{00}/\omega_{01})^{-\omega_{01}}(C_{01}/\omega_{02})^{\omega_{02}}(C_{02}/\omega_{03})^{\omega_{03}}(C_{03}/\omega_{04})^{\omega_{04}}\right]^{-1}$$

$$D(Y) = -1\left[(800/2.5)^{-2.5}(20/0.25)^{0.25}(24/0.75)^{0.75}(30/0.50)^{0.50}\right]^{-1} \tag{15.13}$$

$$= -5,877.12.$$

This implies that the profit is +5,877.12. The primal variables can be found from the primal-dual relationships and the values of the dual objective function and dual variables similar to that in the cost minimization procedure. The equations relating the primal and dual for determining the primal variables are:

$$C_{0t}\prod_{n=1}^{N}x_n^{mtn} = \omega_{0t}\sigma d(\omega). \tag{15.14}$$

Thus,

$$x_1 = \omega_{02}Y/C_{01} = 0.25 \times 5877.12/20 = 73.46 \tag{15.15}$$

$$x_2 = \omega_{03}Y/C_{02} = 0.75 \times 5877.12/24 = 183.66 \tag{15.16}$$

$$x_3 = \omega_{04}Y/C_{03} = 0.50 \times 5877.12/30 = 97.95. \tag{15.17}$$

If the values for x_1, x_2, and x_3 are used in the primal, one obtains:

$$\text{Maximize } \pi = pAx_1^{0.1}x_2^{0.3}x_3^{0.2} - C_{01}x_1 - C_{02}x_2 - C_{03}x_3 \tag{15.1}$$

$$= 800(73.46)^{0.1}(183.66)^{0.3}(97.95)^{0.2} - 20 \times 73.46 - 24 \times 183.66 - 30 \times 97.95$$

$$= 14,692.66 - 1,469.2 - 4,407.84 - 2,938.5$$

$$= 14,692.66 - 8,815.54$$

$$\pi = 5,877.12.$$

Therefore, the values of the primal and dual objective functions are equal. The three cost terms of 1,469.2, 4,407.84, and 2,938.5 are in the same ratio as the cost dual variables of 0.25, 0.75, and 0.50 or 1:3:2. Note that the total revenue is 5/3 of the total cost and that the profits are 2/3 of the total cost which was predicted by the dual variables. Since the dual variables are constants, the ratios will hold for changing inputs constants as long as the equation exponents do not change. However, the primal variables, x_1, x_2, and x_3, are not in the same ratio as the dual variables.

The design equations for the primal variables are more difficult to obtain. The additional primal-dual relationship necessary is:

$$x_1^\alpha x_2^\beta x_3^\gamma = \omega_{01} Y / C_{00}. \tag{15.18}$$

Now using Equations (15.14)–(15.17) and a lot of algebra, one can obtain:

$$x_1 = \left[(\alpha^{1.25} \beta^{0.75} \gamma^{0.50}) / \omega_{01}^{2.5} \right] * \left[C_{00}^{2.50} / (C_{01}^{1.25} C_{02}^{0.75} C_{03}^{0.50}) \right] \tag{15.19}$$

$$x_2 = \left[(\alpha^{0.25} \beta^{1.75} \gamma^{0.50}) / \omega_{01}^{2.5} \right] * \left[C_{00}^{2.50} / (C_{01}^{0.25} C_{02}^{1.75} C_{03}^{0.50}) \right] \tag{15.20}$$

$$x_3 = \left[(\alpha^{0.25} \beta^{0.75} \gamma^{1.50}) / \omega_{01}^{2.5} \right] * \left[C_{00}^{2.50} / (C_{01}^{0.25} C_{02}^{0.75} C_{03}^{1.50}) \right]. \tag{15.21}$$

Since the dual variables are constant as long as the exponents are constant, these design equations can be restated with the current values of α, β, and γ as

$$x_1 = \left[1.0194 * 10^{-2} \right] * \left[C_{00}^{2.50} / (C_{01}^{1.25} C_{02}^{0.75} C_{03}^{0.50}) \right] = 73.46 \tag{15.22}$$

$$x_2 = \left[3.0582 * 10^{-2} \right] * \left[C_{00}^{2.50} / (C_{01}^{0.25} C_{02}^{1.75} C_{03}^{0.50}) \right] = 183.66 \tag{15.23}$$

$$x_3 = \left[2.0389 * 10^{-2} \right] * \left[C_{00}^{2.50} / (C_{01}^{0.25} C_{02}^{0.75} C_{03}^{1.50}) \right] = 97.95. \tag{15.24}$$

These are the same results as obtained by Equations (15.13), (15.14), and (15.15). Similarly, the dual objective function can be reduced to terms of a constant and cost coefficients and converted to profit as:

$$
\begin{aligned}
D(Y) &= +1 \left[(C_{00}/\omega_{01})^{-\omega_{01}} (C_{01}/\omega_{02})^{\omega_{02}} (C_{02}/\omega_{03})^{\omega_{03}} (C_{03}/\omega_{04})^{\omega_{04}} \right]^{-1} \\
&= +1 \left[\left[(\omega_{02})^{\omega_{02}} (\omega_{03})^{\omega_{03}} (\omega_{04})^{\omega_{04}} \right] / (\omega_{01})^{\omega_{01}} \right] \\
&\quad * \left[(C_{00})^{\omega_{01}} / \left[(C_{01})^{\omega_{02}} (C_{02})^{\omega_{03}} (C_{03})^{\omega_{04}} \right] \right] \\
&= \left[\left[(4.0777 * 10^{-2}] * \left[(C_{00})^{2.5} / \left[(C_{01})^{0.25} (C_{02})^{0.75} (C_{03})^{0.50} \right] \right] \right] \right] = 5877.12. \quad (15.25)
\end{aligned}
$$

The design equations, Equations (15.22)–(15.25) can be used for any set of the coefficients to determine the primary variables and the dual objective function. One does not need to resolve the program and these equations permit a rapid method for sensitivity analysis. If the exponents change, the dual variables will change and a new set of equations would need to be developed.

15.2   EVALUATIVE QUESTIONS

1. What is the history of the Cobb-Douglas Production Function?

2. What is the primary difference between the dual objective functions for the minimum cost and maximum profit models?

3. If the market price changes from \$20 to \$25, what is the change in the profit and the amounts of Product 1, Product 2, and Product 3?

4. The profit function is:

$$\text{Maximize } \pi = pAx_1^{0.1}x_2^{0.3}x_3^{0.2} - C_{01}x_1 - C_{02}x_2 - C_{03}x_3^{0.80}.$$

And the values for the primal coefficients are:

$$pA = \$500$$
$$C_{01} = \$40$$
$$C_{02} = \$30$$
$$C_{03} = \$20$$

a. Solve for the dual variables and the dual objective function.

b. Solve for the primal variables and the primal objective function.

c. Show that the cost terms of the primal are in the same ratio as the dual variables.

d. What happens if the exponent on the last term is 0.2 instead of 0.8?

5. The following problem is from the website `http://www.mpri.lsu.edu/textbook/Chapter3-b.htm`

Maximize the function:

$$Y = 3x_1^{0.25} - 3x_1^{1.1}x_2^{0.6} - 115x_2^{-1}x_3^{-1} - 2x_3.$$

a. Solve for the dual variables and the dual objective function.

b. Solve for the primal variables and the primal objective function.

c. Show that the cost terms (negative functions) of the primal are in the same ratio as the dual variables.

(Hint—solve the dual in terms of fractions answers rather than decimal values.)

15.3 REFERENCES

[1] S.-T. Liu, A geometric programming approach to profit maximization, *Applied Mathematics and Computation*, 182, pp. 1093–1097, 2006. 91

[2] M. Keyzer and L. Wesenbeck, Equilibrium selection in games; the mollifier method, *Journal of Mathematical Economics*, 41, pp. 285–301, 2005. 91

[3] `http://www.mpri.lsu.edu/textbook/Chapter3-b.htm`

CHAPTER 16

Chemical Plant Product Profitability Case Study

16.1 MODEL FORMULATION

This chemical plant product profitability problem is from the 1960s and it was presented by Wilde and Beightler [1] in their book, *Foundations of Optimization*, and was a modification of a problem by Wilde's student Ury Passey [2]. It is one of the first demonstrations of the importance of the dual variables in cases of profit consideration where the objective function has both positive and negative terms. The variables for the process considered are:

x_1 = reactor temperature°

x_2 = reactor pressure −atm

x_3 = weight % catalyst

x_4 = weight % product.

The relation between x_4 and x_1 is given as

$$x_4 = 0.100x_1^{0.25}. \tag{16.1}$$

This relation will reduce the number of variables in the solution of the problem and the number of degrees of difficulty. The chemical plant data is presented in Table 16.1.

Table 16.1: Operations and materials for chemical plant

Operation	Material	Cost ($/100 lb processed)	Sales ($/100 lb processed)	Symbol Used
Reactor		$3.18 * 10^{-4} x_1^{1.1} x_2^{0.6}$		C_{01}
Separator		$114.3 * x_2^{-1} x_3^{-1}$		C_{02}
	Catalyst	$2.28 x_3$		C_{03}
	Product		$33.8 x_4 = 33.8 x_1^{0.25}$	C_{04}

16.2 PRIMAL AND DUAL SOLUTIONS

The primal objective function can be expressed as

$$\text{Minimize } Y = +3.18 * 10^{-4} x_1^{1.1} x_2^{0.6} + 114.3 * x_2^{-1} x_3^{-1} + 2.28 * x_3 - 3.38 * x_1^{0.25} \quad (16.2)$$

or in general terms

$$\text{Minimize } Y = C_{01} * x_1^{1.1} x_2^{0.6} + C_{02} * x_2^{-1} x_3^{-1} + C_{03} * x_3 - C_{04} * x_1^{0.25}, \quad (16.3)$$

where $Y = -\pi$ and $\pi = $ profit of project.

The degrees of freedom are obtained by

$$D = T - (N + 1) = 4 - (3 + 1) = 0, \quad (16.4)$$

where

$T = $ number of terms in primal

$N = $ number of variables in primal.

The dual objective function would be:

$$D(Y) = -1 \left[(C_{01}/\omega_{01})^{\omega_{01}} (C_{02}/\omega_{02})^{\omega_{02}} (C_{03}/\omega_{03})^{\omega_{03}} (C_{04}/\omega_{04})^{\omega_{04}} \right]^{-1}. \quad (16.5)$$

For maximization of profits, the signum function $\sigma_{00} = -1$, whereas in the dual objective function for the minimization of costs the signum function $\sigma_{00} = 1$, a positive value. From the coefficients and signs, the signum values for the dual are:

$$\sigma_{01} = \quad 1$$
$$\sigma_{02} = \quad 1$$
$$\sigma_{03} = \quad 1$$
$$\sigma_{04} = -1.$$

The dual formulation would be:

Objective Fctn	$\omega_{01} + \omega_{02} + \omega_{03}$	$- \omega_{04} = -1$ (max)	(16.6)
x_1 terms	$1.1\omega_{01}$	$- 0.25\omega_{04} = \quad 0$	(16.7)
x_2 terms	$0.6\omega_{01} - \omega_{02}$	$= \quad 0$	(16.8)
x_3 terms	$- \omega_{02} + \omega_{03}$	$= \quad 0.$	(16.9)

Solving Equations (16.6) through (16.9) for the dual variables one obtains:

$$\omega_{01} = 5/11$$
$$\omega_{02} = 3/11$$
$$\omega_{03} = 3/11$$
$$\omega_{04} = 2.0.$$

The dual variables represent the relative costs and revenues of the project. The first dual variable represents the reactor cost component, the second dual variable represents the separator cost, the third dual variable represents the catalyst cost, and the fourth dual variable represents the total revenue. Thus, the reactor cost will be 5/11 of the total cost, the separator will be 3/11 of the total cost, and the catalyst cost will be 3/11 of the total cost. The total revenue will be twice the total cost, and thus the profit will be the difference between the total revenue and total cost, and thus equal to the total cost. All this information is known before the primal variables are known and these ratios will hold for any set of cost terms as long as there is no change in the exponents, which would change the dual variables.

Using Equation (16.5), the profit can be calculated as:

$$D(Y) = -1\left[(C_{01}/\omega_{01})^{\omega_{01}}(C_{02}/\omega_{02})^{\omega_{02}}(C_{03}/\omega_{03})^{\omega_{03}}(C_{04}/\omega_{04})^{\omega_{04}}\right]^{-1} \tag{16.5}$$

$$= -1\left[(3.18*10^{-4}/(5/11))^{5/11}(114.3/(3/11))^{3/11}(2.28/(3/11))^{3/11}(3.38/\omega_{04})^2\right]^{-1}$$

$$= -1[0.11933]^{-1}$$

$$= -8.38.$$

Which means that
$$\pi = +\$8.38/100 \text{ lb processed.}$$

This is the total profit and note that none of the primary variables have been determined. The primary variables can be determined from the primal-dual relationships which are given by

$$C_{0t}\prod_{}^{N} x_n^{mtn} = \omega_{0t}\sigma d(\omega). \tag{16.10}$$

The four primal-dual equations are:

$$C_{01}*x_1^{1.1}x_2^{0.6} = \omega_{01}d(\omega) \tag{16.11}$$

$$C_{02}*x_2^{-1}x_3^{-1} = \omega_{02}d(\omega) \tag{16.12}$$

$$C_{03}*x_3 = \omega_{03}d(\omega) \tag{16.13}$$

$$C_{04}*x_1^{0.25} = \omega_{04}d(\omega) \tag{16.14}$$

$$x_1(\text{temperature}) = (\omega_{04}d(\omega)/C_{04})^4 = (2*8.38/3.38)^4 = 605° \tag{16.15}$$

$$x_3(\text{percent catalyst}) = (\omega_{03}d(\omega)/C_{03}) = (3/11*8.38/2.28) = 1.00(\%) \tag{16.16}$$

$$x_2(\text{pressure}) = (C_{02}C_{03}/(\omega_{02}\omega_{03}d(\omega)d(\omega)))$$

$$= ((114.8*2.28)/(3/11*3/11*8.38*8.38)) = 49.9\,\text{atm} \tag{16.17}$$

$$x_4(\text{percent product}) = 0.100x_1^{0.25} = 0.100(\omega_{04}d(\omega)/C_{04})^1$$

$$= 0.100(2*8.38/3.38) = 0.496(\%). \tag{16.18}$$

The primal objective function can now be evaluated by Equation (16.3) as

$$\text{Minimize } Y = C_{01} * x_1^{1.1} x_2^{0.6} + C_{02} * x_2^{-1} x_3^{-1} + C_{03} * x_3 - C_{04} * x_1^{0.25} \tag{16.3}$$

$$= 3.18 * 10^{-4} (605)^{1.1} (49.9)^{0.6} + (114.3)(49.9)^{-1} (1.00)^{-1}$$

$$+ (2.28)(1.00) - (3.38)(605)^{0.25}$$

$$= 3.81 + 2.29 + 2.28 - 16.76$$

$$= -8.38.$$

So then

$$\pi = +8.38,$$

which is the same as the dual objective function calculated by Equation (16.5).

The equations for the primal variables in terms of only the primal constants can be obtained as the dual variables are constants. For example:

$$x_1 = (\omega_{04})^{4(1-\omega_{04})} \left[(\omega_{01})^{\omega_{01}} (\omega_{02})^{\omega_{02}} (\omega_{03})^{\omega_{03}} \right]^4$$

$$* \left[(1/C_{04})^{4(1-\omega_{04})} \left[(C_{01})^{\omega_{01}} (C_{02})^{\omega_{02}} (C_{03})^{\omega_{03}} \right] \right]^{-4}. \tag{16.19}$$

Using the dual variables the expression becomes:

$$x_1 = \left[8.7530 * 10^{-4} \right] * \left[(1/3.38)^{-4} (3.18 * 10^{-4})^{5/11} (114.3)^{3/11} (2.28)^{3/11} \right]^{-4} \tag{16.20}$$

$$= 604.5°$$

$$x_3 = \left[(\omega_{01})^{\omega_{01}} (\omega_{02})^{\omega_{02}} (\omega_{03})^{(1+\omega_{03})} / (\omega_{04}^{\omega_{04}}) \right]$$

$$* \left[((C_{04})^{\omega_{04}}) / \left[(C_{01})^{\omega_{01}} (C_{02})^{\omega_{02}} \left[(C_{03})^{(1+\omega_{03})} \right] \right] \right] \tag{16.21}$$

$$x_3 = \left[2.345512 * 10^{-2} \right] * \left[((C_{04})^{\omega_{04}}) / \left[(C_{01})^{\omega_{01}} (C_{02})^{\omega_{02}} \left[(C_{03})^{(1+\omega_{03})} \right] \right] \right] \tag{16.22}$$

$$= \left[2.345512 * 10^{-2} \right] * \left[(3.38)^2 / \left[(3.18 * 10^{-4})^{5/11} (114.3)^{3/11} \left[(2.28)^{(1+3/11)} \right] \right] \right]$$

$$= 1.00\%$$

$$x_2 = \left[(1/(\omega_{02}\omega_{03})) ((\omega_{04})^{\omega_{04}}) / \left[(\omega_{01})^{\omega_{01}} (\omega_{02})^{\omega_{02}} (\omega_{03})^{\omega_{03}} \right] \right]^2$$

$$* \left[(C_{02}C_{03}) * \left[((C_{01})^{\omega_{01}} (C_{02})^{\omega_{02}} (C_{03})^{\omega_{03}}) / (C_{04})^{\omega_{04}} \right] \right]^2 \tag{16.23}$$

$$x_2 = [1817.7] * \left[(C_{02}C_{03}) * \left[((C_{01})^{\omega_{01}} (C_{02})^{\omega_{02}} \left[(C_{03})^{\omega_{03}} \right]) / (C_{04})^{\omega_{04}} \right] \right]^2 \tag{16.24}$$

$$= [1817.7] * \left[(114.3 * 2.28) * \left[\left((3.18 * 10^{-4})^{5/11} (114.3)^{3/11} \left[(2.28)^{3/11} \right] \right) / (3.38)^2 \right] \right]^2$$

$$= [1817.7] * [260.604] * \left[1.05324 * 10^{-4} \right]$$

$$= 49.9 \, \text{Atm.}$$

These design equations no longer have the dual objective function as a component and thus the primal variables can be obtained from the primal constant values and the dual variables which are constant as long as the exponents do not change.

16.3   EVALUATIVE QUESTIONS

1. Show that the four primal terms are in the same ratio as the dual variables for the example problem.

2. Find the Equation for x_4 in terms of the constant coefficients of the primal and let $C_{05} = 0.100$ and combine the dual parts into a single constant similar to that of Equation (16.22).

3. Change the catalyst cost from 2.28 to 4.56 and what are the dual variables, dual objective function, primal variables, primal objective function, and cost ratios?

16.4   REFERENCES

[1] D. J. Wilde and C. S. Beightler, *Foundations of Optimization*, Prentice-Hall, Englewood Cliffs, NJ, 1967. 97

[2] U. Passy and D. J. Wilde, Generalized polynomial optimization, *Stanford Chemical Engineering Report*, 1966. 97

PART IV

Geometric Programming Applications with Positive Degrees of Difficulty

CHAPTER 17

Journal Bearing Design Case Study

17.1 ISSUES WITH POSITIVE DEGREES OF DIFFICULTY PROBLEMS

Problems with one positive degree of difficulty are more difficult solve as the additional equation typically is not linear in the dual, and the design equations are also more difficult to obtain. Problems with multiple degrees of difficulty are very difficult to solve and thus it is very difficult to obtain design equations for those problems. They typically require advanced search techniques in order to obtain results, and more research is required to obtain better techniques to obtain the design equations.

Some basic techniques for obtaining additional equations and a solution include the substitution approach, dimensional analysis, the constrained derivative approach, and the condensation of terms approach are presented in the following chapters. The substitution approach and dimensional analysis methods typically add one additional equation to reduce the degree of difficulty, but sometimes more than one additional equation can be obtained. The constrained derivative approach usually adds one additional equation by differentiation which reduces the degree of difficulty by one. The condensation of terms approach combines two terms into one term and thus also reduces the degrees of difficulty, but some accuracy is sacrificed. The condensation of terms can possibly reduce more than one term, but the solution will lose more accuracy. These methods will be illustrated in the following chapters.

In the cases of zero degrees of difficulty, the dual variables were independent of the primal constants. When the degrees of difficulty are positive, the dual variables are often functions of the primal constants, and thus the dual variables would change when the primal constants change. Thus, the ratios between the terms would not be independent of the primal constants. When there is only one degree of difficulty, it often is easy to obtain the design equations. But with multiple degrees of difficulty it is much more difficult to obtain design equations.

17.2 JOURNAL BEARING CASE STUDY

An interesting problem with one degree of difficulty is a journal bearing design problem presented by Beightler, Lo, and Bylander [1]. The objective was to minimize the cost (P), and the variables were the half-length of the bearing (L) and the radius of the journal (R). The objective function presented and the constants in the problem are those presented in the original paper and the derivations of the constants were not presented in detailed. The original problem [1] was solved by determining upper and lower bounds to the solution and search techniques in the reduced variable range. The solution now presented solves the problem directly using the additional equation without needing to use search techniques, and design equations are obtained for the variables. The substitution approach and the dimensional analysis approach are demonstrated for obtaining the additional equation.

17.3   PRIMAL AND DUAL FORMULATION OF JOURNAL BEARING DESIGN

The generalized primal problem was:

$$\text{Minimize } P = C_{01} R^3 L^{-2} + C_{02} R^{-1} + C_{03} RL^{-3} \tag{17.1}$$
$$\text{Subject to:} \quad C_{11} * R^{-1} * L^3 \leq 1, \tag{17.2}$$

where

$P = \text{cost (\$)}$
$R = \text{radius of the journal (in)}$
$L = \text{half-length of the bearing (in)}$
$C_{01} = 0.44 \text{ (for example problem)}$
$C_{02} = 10 \text{ (for example problem)}$
$C_{03} = 0.592 \text{ (for example problem)}$
$C_{11} = 8.62 \text{ (for example problem)}.$

Figure 17.1 is a sketch illustrating the variables for the problem.

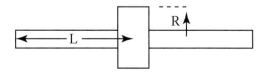

Figure 17.1: Journal bearing parameters of half-length and radius.

From the coefficients and signs, the signum values for the dual from Equations (17.1) and (17.2) are:

$$\sigma_{01} = 1$$
$$\sigma_{02} = 1$$
$$\sigma_{03} = 1$$
$$\sigma_{11} = 1$$
$$\sigma_1 \ = 1.$$

The dual problem formulation is:

Objective Function	$\omega_{01} + \omega_{02} + \omega_{03}$ $= 1$	(17.3)
R terms	$3\omega_{01} - \omega_{02} + \omega_{03} - \omega_{11} = 0$	(17.4)
L terms	$-2\omega_{01} \quad -3\omega_{03} + 3\omega_{11} = 0.$	(17.5)

From the constraint equation there is only one term, so:

$$\omega_{10} = \omega_{11}. \tag{17.6}$$

This adds one additional equation but also one additional term, so the degrees of difficulty are equal to:

$$D = T - (N + 1) = 4 - (2 + 1) = 1 \geq 0. \tag{17.7}$$

The dual has more variables than equations, and thus another equation is needed to solve for the dual variables. The relationships between the primal and dual variables will be used to determine an additional equation, and the equation typically is non-linear. The approach presented here is the "substitution approach" to obtain the additional equation needed. The "dimensional analysis approach" will be presented later in this chapter. The relationships between the primal and dual are:

$$C_{01} R^3 L^{-2} = \omega_{01} P \tag{17.8}$$
$$C_{02} R^{-1} = \omega_{02} P \tag{17.9}$$
$$C_{03} R L^{-3} = \omega_{03} P \tag{17.10}$$
$$C_{11} R^{-1} L^3 = (\omega_{11}/\omega_{10}). \tag{17.11}$$

Since $\omega_{10} = \omega_{11}$, Equation (17.11) can be used to relate the primal variables, that is:

$$R = C_{11} L^3. \tag{17.12}$$

Using Equation (17.12) in Equation (17.9), one obtains:

$$P = C_{02}/(\omega_{02} R)$$
$$= \left[C_{02}/(C_{11}\omega_{02} L^3) \right]. \tag{17.13}$$

Using Equation (17.10) with Equations (17.12) and (17.13), one obtains after reducing terms:

$$L^3 = (C_{03}R)/(\omega_{03}P)$$
$$L^3 = \left[(C_{02}/(C_{03}C_{11}^2)) * (\omega_{03}/\omega_{02})\right]. \tag{17.14}$$

Now using Equation (17.8) and the values for R and L, one obtains:

$$\left(C_{01}C_{11}^4 L^{10}\right)/C_{02} = \omega_{01}/\omega_{02}. \tag{17.15}$$

Now using Equation (17.14) in Equation (17.15) and reducing it one can obtain:

$$\left(C_{01}C_{02}^{7/3}\right)/\left(C_{03}^{10/3}C_{11}^{8/3}\right) = \left(\omega_{01}\omega_{02}^{7/3}/\omega_{03}^{10/3}\right) \tag{17.16}$$

or the form of

$$\left(C_{01}^3 C_{02}^7\right)/\left(C_{03}^{10}C_{11}^8\right) = \left(\omega_{01}^3 \omega_{02}^7/\omega_{03}^{10}\right). \tag{17.17}$$

Now using Equations (17.3) to (17.5) to solve for the dual variables in terms of ω_{02}, one obtains:

$$\omega_{01} = (3/7)\omega_{02} \tag{17.18}$$
$$\omega_{03} = 1 - (10/7) * \omega_{02} \tag{17.19}$$
$$\omega_{11} = 1 - (8/7) * \omega_{02}. \tag{17.20}$$

Using Equations (17.18) to (17.20) in Equation (17.17), one can obtain:

$$(3/7\omega_{02})^3 (\omega_{02})^7/[1 - ((10/7) * \omega_{02})]^{10} = \left(C_{01}^3 C_{02}^7\right)/\left(C_{03}^{10}C_{11}^8\right) = A \quad \text{or}$$
$$(\omega_{02}/(1 - ((10/7)\omega_{02})) = \left[A * (7/3)^3\right]^{1/10} = B \quad \text{or}$$
$$\omega_{02} = (7B/(7 + 10B)). \tag{17.21}$$

Thus, the remaining dual variables can be solved for as:

$$\omega_{01} = 3B/(7 + 10B) \tag{17.22}$$
$$\omega_{03} = 7/(7 + 10B) \tag{17.23}$$
$$\omega_{11} = (7 + 2B)/(7 + 10B). \tag{17.24}$$

Using Equations (17.10) and (17.12)

$$
\begin{aligned}
P &= (C_{03}/\omega_{03})RL^{-3}\\
&= (C_{03}/\omega_{03})\left(C_{11}L^{3}\right)L^{-3}\\
&= C_{11}C_{03}/\omega_{03}\\
&= (C_{11}C_{03})(1 + (10/7)B)\\
&= (C_{11}C_{03})\left(1 + (10/7)\left[(7/3)^{3}A\right]^{1/10}\right)\\
&= (C_{11}C_{03})(1 + (10/7)(7/3)^{3/10}\left[(C_{01}^{3}C_{02}^{7})/(C_{03}^{10}C_{11}^{8})\right]^{1/10}\\
&= C_{11}C_{03} + C_{11}C_{03}(10/7)(7/3)^{3/10}\left[(C_{01}^{3}C_{02}^{7})/(C_{03}^{10}C_{11}^{8})\right]^{1/10}\\
&= C_{11}C_{03} + (10/7)[((7/3)(C_{01})]^{3/10} * (C_{02})^{7/10} * (C_{11})^{2/10}. \quad (17.25)
\end{aligned}
$$

Using Equation (17.9) to solve for R one obtains:

$$
\begin{aligned}
R &= C_{02}/(\omega_{02}P)\\
&= C_{02}/(7B/(7 + 10B)) * (C_{11}C_{03})(1 + (10/7)B)\\
&= C_{02}/(7B/(7 + 10B)) * (C_{11}C_{03})(7 + (10B)7)\\
&= (C_{02}/(C_{11}C_{03}))/B\\
&= (C_{02}/(C_{11}C_{03}))/\left[(7/3)^{3} * C_{01}^{3}C_{02}^{7}/(C_{03}^{10}C_{11}^{8})\right]^{1/10}\\
&= [(3/7) * (C_{02}/C_{01})]^{3/10}C_{11}^{-2/10}. \quad (17.26)
\end{aligned}
$$

Using equation (17.10) to solve for L one obtains:

$$
\begin{aligned}
L &= [C_{03}R/(\omega_{03}P)]^{1/3}\\
&= \left[C_{03} * [(3/7) * (C_{02}/C_{01})]^{3/10} C_{11}^{-2/10}\right]/[\omega_{03} * C_{11}C_{03}/\omega_{03}]\\
&= [(3/7)(C_{02}/C_{01})]^{1/10} * C_{11}^{-4/10}. \quad (17.27)
\end{aligned}
$$

The equations for P, R, and L are general design equations, but are rather complex equations compared to the previous problems illustrated. The solution was based upon determining an additional equation from the primal-dual relationships, which was highly non-linear and resulted in rather complex expressions for the variables. The additional equation, along with the dual variables, was used in the equations relating the primal and dual to determine the final expressions for the variables. This frequently happens when the degrees of difficulty are greater than zero.

For this particular example problem where $C_{01} = 0.44$, $C_{02} = 10$, $C_{03} = 0.592$, and $C_{11} = 8.62$, the value for A and B are:

$$
A = \left(C_{01}^{3}C_{02}^{7}\right)/\left(C_{03}^{10}C_{11}^{8}\right) = \left[(0.44)^{3}(10)^{7}\right]/\left[(0.592)^{10}(8.62)^{8}\right] = 5.285 \quad (17.28)
$$

$$
B = \left[A(7/3)^{3}\right]^{1/10} = \left[5.285(7/3)^{3}\right]^{1/10} = 1.523. \quad (17.29)
$$

Now using the equations for the dual variables, Equations (17.21) to (17.24), one obtains

$$\omega_{02} = 7B/(7 + 10B) = 0.480$$
$$\omega_{01} = 3B/(7 + 10B) = 0.205$$
$$\omega_{03} = 7/(7 + 10B) = 0.315$$
$$\omega_{11} = (7 + 2B)/(7 + 10B) = 0.452.$$

The dual variables imply that the first term of the variable cost of the primal will be 20.5% of the total cost, the second term of the primal will be 48% of the total cost, and the third term of the primal will be 31.5% of the total cost.

From Equation (17.25) the value of P can be found as:

$$P = C_{11}C_{03} + (10/7)[((7/3)(C_{01})]^{3/10} * (C_{02})^{7/10} * (C_{11})^{2/10}$$
$$= (8.62)(0.592) + (10/7)[((7/3)(0.44)]^{0.3}(10)^{0.7}(8.62)^{0.2}$$
$$= 5.10 + 11.10$$
$$= \$16.2.$$

The primal variables can be determined from Equations (17.26) and (17.27) as:

$$R = [(3/7)(C_{02}/C_{01})]^{3/10}C_{11}^{-2/10}$$
$$= [(3/7)(10/0.44]^{3/10}8.62^{-2/10}$$
$$= 1.29 \text{ in}$$
$$L = [(3/7)(C_{02}/C_{01})]^{1/10}C_{11}^{-4/10}$$
$$= [(3/7)(10/0.44)]^{1/10}8.62^{-4/10}$$
$$= 0.530 \text{ in.}$$

Now if the values of R and L are used in Equation (17.1) for evaluating the primal, one obtains:

$$P = C_{01}R^3L^{-2} + C_{02}R^{-1} + C_{03}RL^{-3}$$
$$= 0.44(1.29)^3(0.53)^{-2} + 10(1.29)^{-1} + 0.592(1.29)(0.53)^{-3}$$
$$= 3.363 + 7.752 + 5.130$$
$$= \$16.245.$$

As in the previous case studies, the values of the primal and dual objective functions are equivalent. Note that the terms of the primal are in the same ratio as that predicted by the dual variables, although some minor rounding has occurred in the calculations.

17.4 DIMENSIONAL ANALYSIS TECHNIQUE FOR ADDITIONAL EQUATION

It was difficult to determine the additional equation by repeated substitution, and other methods can be used. One method is the technique of dimensional analysis and this will be demonstrated to obtain Equation (17.17). The dimensional analysis approach sets up the primal dual relations of Equations (17.8)–(17.11) and setting the primal variables on one side and the dual variables, constants, and objective function on the other side and giving the terms variable exponents as illustrated in Equation (17.30).

$$\left(R^3 L^{-2}\right)^A \left(R^{-1}\right)^B \left(RL^{-3}\right)^C \left(R^{-1}L^3\right)^D = 1$$
$$= (\omega_{01} P/C_{01})^A (\omega_{02} P/C_{02})^B (\omega_{03} P/C_{03})^C (1/C_{11})^D. \tag{17.30}$$

One must balance the exponents to remove the primal variables (R, L) and the dual objective function (P). This is done by:

$$R \text{ values} \quad 3A - B + C - D = 0 \tag{17.31}$$
$$L \text{ values} \quad -2A \qquad -3C + 3D = 0 \tag{17.32}$$
$$P \text{ values} \quad A + B + C \qquad = 0. \tag{17.33}$$

There are four variables and three equations, so one will find three variables in terms of the fourth variable. If one adds three times Equation (17.31) to Equation (17.32), one obtains:

$$A = 3/7B. \tag{17.34}$$

Using Equations (17.34) and (17.33), one can determine that:

$$B = -7/10C, \tag{17.35}$$

which results in:

$$A = -3/10C. \tag{17.36}$$

Using Equations (17.31), (17.35), and (17.36), one obtains that

$$D = 8/10C. \tag{17.37}$$

If one lets $C = 10$, the $A = -3$, $B = -7$, and $D = 8$. Using these values in Equation (17.30), one can obtain Equation (17.17), that is:

$$\left(C_{01}^3 C_{02}^7\right) / \left(C_{03}^{10} C_{11}^8\right) = \left(\omega_{01}^3 \omega_{02}^7 / \omega_{03}^{10}\right). \tag{17.17}$$

Another solution method that is often used is the constrained derivative approach. This method has the dual equations rearranged in terms of one unknown dual variable and these are substituted into the dual objective function. The objective function is set into logarithmic form and differentiated with respect to the unknown dual variable, set to zero, and then solved for the unknown dual variable. The solved dual variable is used in the dual equations to obtain the values of the other dual variables. This technique is demonstrated in Chapters 21, 22, and 23.

17.5   EVALUATIVE QUESTIONS

1. Use the values of $C_{01} = 0.54, C_{02} = 10, C_{03} = 0.65$, and $C_{11} = 13.00$, determine the values of A and B, of the dual variables and the value of the objective function. Also determine the values of L and R and use these to determine P.

2. Determine the sensitivity of the objective function and the primal variables of R and L by changing one of the constants by 20% (such as C_{01}).

3. Use dimensional analysis to develop the additional equation for the dual variables (ω values) in terms of the constants (C values) from the following data where D and H are primal variables and Y is the dual objective function.

$$C_{01} D^2 H = \omega_{01} Y$$
$$C_{02} D^3 = \omega_{02} Y$$
$$C_{11} D^{-1} = \omega_{11}/\omega_{10}$$
$$C_{12} H^{-1} = \omega_{12}/\omega_{10}$$
$$C_{13} D H^{-1} = \omega_{13}/\omega_{10}$$

[Answer $(C_{02}/C_{01})(C_{11}/C_{12}) = (\omega_{02}/\omega_{01})(\omega_{11}/\omega_{12})$].

17.6   REFERENCES

[1] C. S. Beightler, T-C. Lo, and H. G. Bylander, Optimal design by geometric programming, ASME, *Journal of Engineering for Industry*, pp. 191–196, 1970. 105, 106

[2] R. C. Creese, A primal-dual solution procedure for geometric programming, ASME, *Journal of Mechanical Design*, (also as Paper No. 79-DET-78), 1979.

CHAPTER 18

Multistory Building Design with a Variable Number of Floors Case Study

18.1 INTRODUCTION

Chapter 6, "The Building Design Case Study," considered a building with a fixed number of floors. The number of floors was fixed at three—the ground floor and two suspended slab floors. This new model allows the number of floors to vary and the solution will typically be a non-integer which then must be evaluated at the two boundary integers. The new model also includes the purchase of the land as the number floors increase, the amount of land required will decrease for the same amount of floor space. The room height will be a specified value to prevent unreasonable values as obtained in the basic model in Chapter 6. The painted area considered will be the total wall length, and windows can be considered as a fraction of the total wall surface.

18.2 PROBLEM FORMULATION

A new building is to be constructed with a volume of 30,000 cubic meters. The problem is to determine the number of floors and what the dimensions of the building should be. The length, width, and height of the building are variable, the floor height is specified at 3 meters and the total number of floors is variable, consisting of a single ground floor (slab-on-grade) and a variable number of raised floors (suspended slabs). The number of floors must be an integer. The dimensions of the building and the costs are:

Input

$V = 30,000 \text{ m}^3 = $ volume of building

$H = $ gross height of room $= 3 \text{ m}$

$A = $ floor area $= V/H = 30,000/3 = 10,000 \text{ m}^2$

$L = $ length of outside walls

$W = $ width of outside walls of the building $= L$ (for a square building)

$N = $ total floors which includes the ground floor

$C_L = 50 \ \$/m^2 = $ cost of land

$C_G = 150/m^2 = $ cost of ground floor

$C_R = 70/m^2 = $ cost of roof

$C_F = 100/m^2 = $ cost of intermediate (suspended slab) floor

$C_W = 180/m^2 = $ cost of walls

$C_T = $ total cost.

The primal objective function would be:

$$C_T = \text{Land Cost} + \text{Ground Floor Cost}$$
$$+ \text{Roof Cost} + \text{Suspended Floor Cost} + \text{Wall Cost} \tag{18.1}$$
$$C_T = C_L * L^2 + C_G * L^2 + C_R * L^2 + (N-1) * C_F * L^2 + 4 \times H * L * N * C_w, \tag{18.2}$$

and the primal objective function would be:

$$C_T = C_{01} * L^2 + N * C_{02} * L^2 + N * C_{03} L, \tag{18.3}$$

where

$$C_{01} = C_L + C_G + C_R - C_F = 50 + 150 + 70 - 100 = 170 \tag{18.4}$$
$$C_{02} = C_F = 100 \tag{18.5}$$
$$C_{03} = 4 \times 3 (= H) \times C_w = 4 \times 3 \times 180 = 2,160, \tag{18.6}$$

and C_{01} must be positive.

Subject to the total volume constraint:

$$V = A * H = N * L^2 * H = N * L^2 * 3 >= 30,000 \tag{18.7}$$

or

$$N \times L^2 >= 10,000, \tag{18.8}$$

which implies the total floor area is 10,000 m^2 area

$$10,000/ \left(N * L^2 \right) <= 1, \tag{18.9}$$

or in geometric programming constraint form:

$$C_{11} * N^{-1} * L^{-2} <= 1, \tag{18.10}$$

where $C_{11} = 10,000$.

If one examines the primal objective function [Equation (18.3)] and the constraint [Equation (18.10)], there are four terms and two variables, so the degree of difficulty is:

$$D = T - (N + 1) \tag{18.11}$$
$$T = \text{number of terms} = 4$$
$$N = \text{number of variables} = 2$$
$$D = 4 - (2 + 1) = 1. \tag{18.12}$$

Since there is only one degree of difficulty, one will search for an additional equation. Thus, the dual objective function is:

$$Y = d(\omega) = \sigma_{00} * [(C_{01} * \omega_{00}/\omega_{01})^{\sigma_{01}\omega_{01}} * (C_{02} * \omega_{00}/\omega_{02})^{\sigma_{02}\omega_{02}}$$
$$*(C_{03} * \omega_{00}/\omega_{03})^{\sigma_{03}\omega_{03}} * (C_{11} * \omega_{10}/\omega_{11})^{\sigma_{11}\omega_{11}}]^{\sigma_0}, \tag{18.13}$$

where

$$\sigma_{00} = 1 \,(\text{minimization}) \text{ and } \sigma_{01} = \sigma_{02} = \sigma_{03} = 1 \quad \text{and} \quad \sigma_{11} = 1,$$

and since only one term is in the constraint, $\omega_{10} = \omega_{11}$ and thus $\omega_{10}/\omega_{11} = 1$. The dual problem can now be formulated as:

$$\text{Obj} \qquad \omega_{01} + \quad \omega_{02}+\omega_{03} \qquad = 1 \tag{18.14}$$
$$L \text{ terms} \qquad 2\omega_{01} + 2\omega_{02}+\omega_{03} - 2\omega_{11}= 0 \tag{18.15}$$
$$N \text{ terms} \qquad + \quad \omega_{02}+\omega_{03} - \quad \omega_{11}= 0. \tag{18.16}$$

If one takes 2 times Equation (18.14) and subtracts Equation (18.15), one can obtain

$$\omega_{03} = 2 \times (1 - \omega_{11}). \tag{18.17}$$

Using Equation (18.17) in Equation (18.16)

$$\omega_{02} = 3\omega_{11} - 2. \tag{18.18}$$

Using Equation (18.16) in Equation (18.14) one obtains

$$\omega_{01} = 1 - \omega_{11}. \tag{18.19}$$

The primal-dual relationships are used to determine the additional equation using dimensional analysis. The primal-dual equations for the objective function and constraint terms are:

$$C_{ot} \prod_{n=1}^{N} X_n^{a_{mtn}} = \omega_{ot}\sigma d(\omega) \qquad t = 1, \dots, T_o, \tag{18.20}$$

and

$$C_{mt} \prod_{n=1}^{N} X_n^{a_{mtn}} = \omega_{mt}/\omega_{mo} \qquad t = 1, \ldots, T_o \quad \text{and} \quad m = 1, \ldots, M. \tag{18.21}$$

The dimensional analysis approach sets up the primal dual relations of Equations (18.20) and (18.21) setting the primal variables on one side and the dual variables, constants, and objective function on the other side and giving the terms variable exponents (A, B, C, D) as illustrated in Equation (18.22).

$$\left(L^2\right)^A \left(NL^2\right)^B (NL)^C \left(N^{-1}L^{-2}\right)^D = 1$$
$$= (\omega_{01}Y/C_{01})^A (\omega_{02}Y/C_{02})^B (\omega_{03}Y/C_{03})^C (\omega_{10}/\omega_{11}/C_{11})^D. \tag{18.22}$$

Since $\omega_{10} = \omega_{11}$, the last term can be reduced and the equation becomes

$$\left(L^2\right)^A \left(NL^2\right)^B (NL)^C \left(N^{-1}L^{-2}\right)^D = 1$$
$$= (\omega_{01}Y/C_{01})^A (\omega_{02}Y/C_{02})^B (\omega_{03}Y/C_{03})^C (1/C_{11})^D. \tag{18.22}$$

Thus, examining the primal variables N and L and the objective function Y, one obtains the following relationships:

Y term exponents	$A + B + C = 0$	(18.23)
L term exponents	$2A + 2B + C - 2D = 0$	(18.24)
N term exponents	$B + C - D = 0.$	(18.25)

If one takes 2 times Equation (18.23) minus Equation (18.24), one obtains that:

$$C = -2D. \tag{18.26}$$

If one uses Equation (18.26) in Equation (18.25), one obtains that:

$$B = 3D. \tag{18.27}$$

Using the results of Equations (18.26) and (18.27) in Equation (18.23) one obtains that

$$A = -D. \tag{18.28}$$

If one lets $D = -1$, then $A = 1$, $B = -3$, and $C = 2$ and then the right-hand side of Equation (18.22) becomes

$$(\omega_{01}Y/C_{01})^1 (\omega_{02}Y/C_{02})^{-3} (\omega_{03}Y/C_{03})^2 (1/C_{11})^{-1} = 1, \tag{18.29}$$

which results in

$$(\omega_{01})^1(\omega_{02})^{-3}(\omega_{03})^2 = (C_{01})^1(C_{02})^{-3}(C_{03})^2(C_{11})^{-1}, \tag{18.30}$$

or

$$\omega_{01}\omega_{02}^{-3}\omega_{03}^2 = C_{01}^1 C_{02}^{-3} C_{03}^2 C_{11}^{-1}$$
$$= K = (170) \times (100)^{-3} \times (2{,}160)^2 \times (10{,}000)^{-1} = 0.0793152 = K. \tag{18.31}$$

Substituting the dual variables in terms of ω_{11}

$$(1 - \omega_{11})(2 \times (1 - \omega_{11}))^2(3\omega_{11} - 2)^{-3} = (4)(1 - \omega_{11})^3/(3\omega_{11} - 2)^3$$
$$= 0.0793152 = K, \tag{18.32}$$

or

$$(1 - \omega_{11})^3/(3\omega_{11} - 2)^3 = [(1 - \omega_{11})/(3\omega_{11} - 2)]^3$$
$$= 0.0793152/4 = 0.0198288 = K/4, \tag{18.33}$$

taking the cube root, one obtains

$$(1 - \omega_{11})/(3\omega_{11} - 2) = 0.27066 = Z = (K/4)^{(1/3)}, \tag{18.34}$$

solving for ω_{11}

$$\omega_{11} = (1 + 2 \times Z)/(1 + 3 \times Z) = (1 + 2 \times 0.27066)/(1 + 3 \times 0.27066) = 0.8506, \tag{18.35}$$

using $\omega_{11} = 0.85$ and thus

$$\omega_{01} = 1 - \omega_{11} = 1 - 0.85 = 0.15 \tag{18.36}$$
$$\omega_{02} = 3\omega_{11} - 2 = 3 \times 0.85 - 2 = 0.55 \tag{18.37}$$
$$\omega_{03} = 2(1 - \omega_{11}) = 2 \times (1 - 0.85) = 0.30. \tag{18.38}$$

From the primal-dual equations

$$C_{01}L^2 = \omega_{01}Y \quad \text{and} \quad C_{02}NL^2 = \omega_{02}Y,$$

one obtains

$$N = C_{01}\omega_{02}/C_{02}\omega_{01} = 170 \times 0.55/(100 \times 0.15)$$
$$= (11/3) * (C_{01}/C_{02}) = 6.23 \text{ floors} \tag{18.39}$$

and

$$L = C_{03}\omega_{02}/C_{02}\omega_{03} = 2{,}160 \times 0.55/(100 \times 0.30)$$
$$= (11/6) * (C_{03}/C_{02}) = 39.6 \text{ m.} \tag{18.40}$$

Since N must be an integer, one evaluates the cases for 6 and 7 floors. Using the constraint equation and solving for L, one obtains:

$$L = (10,000/N)^{0.5}. \tag{18.41}$$

For $N = 6$

$$L = (10,000/6)^{0.5} = 40.8$$
$$C(6\,\text{stories}) = 170 \times 40.8^2 + 100 \times 6 \times 40.8^2 + 2,160 \times 40.8 \times 6$$
$$= 1,810,547$$
$$= 282,989 + 998,784 + 533,664 = 1,810,547. \tag{18.42}$$

These values will not be in the exact same ratio as the dual variables of 0.15:0.55:0.30 as they are for the optimal design of 6.23 floors, but they will be near those values.

For $N = 7$

$$L = (10,000/7)^{0.5} = 37.8,$$
$$C(7\,\text{stories}) = 170 \times 37.8^2 + 100 \times 7 \times 37.8^2 + 2,160 \times 37.8 \times 7$$
$$= 1,814.626. \tag{18.43}$$

Thus, the minimum cost would be obtained for a structure with 6 floors. The dual objective function would be for the 6.23 floors and the primal objective function is necessary to evaluate the two adjacent integer cases of 6 floors and 7 floors. Note that the dual variables are a function of the input values (The K and Z values used) and if the constants change, the dual variables would also change.

In this example one term was removed by combining the negative cost item with the other positive terms of C_{01}. This reduced the degree of difficulty and made the solution easier. The requirement of an integer requirement was also illustrated by this problem.

18.3   EVALUATIVE QUESTIONS

1. What is the solution if the land cost was \$500/m²?

2. The management wants to change the example problem from a square shape to a rectangle where the width is 1/2 the length. Reformulate the problem and determine the solution and what are the new dimensions and cost.

3. If windows are required on each floor as 20% of the wall area at a cost of \$500/m², use an adjusted expression for the wall cost and determine the new solution.

18.4 REFERENCES

[1] M. Gates and A. Scarpa, Optimum configuration of buildings and site utilization, *Cost Engineering*, Vol. 24, No. 6, pp. 333–346, 1982.

CHAPTER 19

Metal Casting Cylindrical Side Riser With Hemispherical Top Design Case Study

19.1 INTRODUCTION

The sphere is the shape which gives the longest solidification time, so a riser with a hemispheric shaped top should be a more efficient riser than a cylindrical riser and thus be more economical. However, the problem of designing this type of riser is more complex and it has two degrees of difficulty. The case study considered is a side riser with a hemispherical top and a cylindrical bottom for ease of molding and to provide a better connection to the casting. The top hemispherical cap has the same diameter (D) as the cylinder and is illustrated in Figure 19.1.

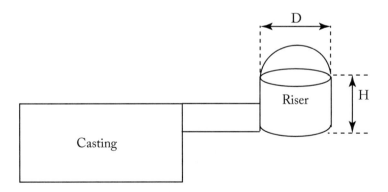

Figure 19.1: Cylindrical side riser with hemispherical top.

The diameter of the hemispherical top has the same diameter of the cylinder portion of the riser. The height of the hemispherical top is equal to the radius of the riser.

19.2   PROBLEM FORMULATION

The volume of the riser is the sum of the cylinder part and the hemisphere part and can be written as:

$$V = \text{cylindrical part} + \text{hemispherical part}$$

$$V = \pi D^2 H/4 + \pi D^3/12 \tag{19.1}$$

$$SA = \pi D^2/4 + \pi DH + \pi D^2/2$$

$$= 3/4\pi D^2 + \pi DH. \tag{19.2}$$

The constraint for riser design is Chvorinov's Rule, which is

$$t = K(V/SA)^2, \tag{19.3}$$

where

$t =$ solidification time (minutes or seconds)

$K =$ solidification constant for molding material (minutes/in^2 or seconds/cm^2)

$V =$ volume (in^3 or cm^3)

$SA =$ cooling surface area (in^2 or cm^2).

This results in the relation:

$$(V/SA) \geq M_c = K, \tag{19.4}$$

where $M_c =$ the modulus of the casting (a constant K for a particular casting).

Thus, using the cooling surface area (SA) and volume (V) expressions, Equation (19.4) can be rewritten as:

$$V/SA = \left(\pi D^2 H/4 + \pi D^3/12\right) / \left(3/4\pi D^2 + \pi DH\right) \geq K$$

$$= \left(\pi D^2/12\right) * (3H + D)/(\pi D/4) * (3D + 4H) \geq K$$

$$= D * (3H + D)/[3 * (3D + 4H)] \geq K. \tag{19.5}$$

Rearranging the equation in the less than equal form results in:

$$4KD^{-1} + 3KH^{-1} - (1/3)DH^{-1} \leq 1. \tag{19.6}$$

Thus, the primal form of the problem can be stated as:

$$\text{Min } V = \pi D^2 H/4 + \pi D^3/12. \tag{19.7}$$

Subject to:

$$4KD^{-1} + 3KH^{-1} - (1/3)DH^{-1} \leq 1. \tag{19.8}$$

From the coefficients and signs, the signum values for the dual are:

$$\sigma_{01} = 1$$
$$\sigma_{02} = 1$$
$$\sigma_{11} = 1$$
$$\sigma_{12} = 1$$
$$\sigma_{13} = -1$$
$$\sigma_{1} = 1.$$

The dual problem formulation is:

Objective Function	$\omega_{01} + \omega_{02}$	$= 1$	(19.9)
D terms	$2\omega_{01} + 3\omega_{02} - \omega_{11}$ $\quad -\omega_{13} = 0$		(19.10)
H terms	$\omega_{01} \qquad\qquad -\omega_{12} + \omega_{13} = 0.$		(19.11)

The degrees of difficulty are equal to:

$$D = T - (N + 1) = 5 - (2 + 1) = 2. \tag{19.12}$$

The dual variables cannot be determined directly as the degrees of difficulty are 2, that is, there are two more variables than there are equations, and additional equations will be sought. Using the linearity inequality equation,

$$\omega_{10} = \omega_{mt} = \sigma_m \sum \sigma_{mt}\omega_{mt} = (1) * (1 * \omega_{11} + 1 * \omega_{12} + (-1) * \omega_{13})$$
$$\omega_{10} = \omega_{11} + \omega_{12} - \omega_{13}. \tag{19.13}$$

The objective function can be found using the dual expression:

$$V = d(\omega) = \sigma \left[\prod_{m=0}^{M} \prod_{t=1}^{T_m} (C_{mt}\omega_{mo}/\omega_{mt})^{\sigma_{mt}\omega_{mt}} \right]^{\sigma} \tag{19.14}$$

$$V = 1 \left[\left[\{(\pi/4) * 1/\omega_{01}\}^{(1*\omega_{01})} \right] * \left[\{(\pi/12) * \omega_{02}\}^{(1*\omega_{02})} \right] \right.$$
$$* \left[\{(4K * \omega_{10}/\omega_{11}\}^{(1*\omega_{10})} \right]$$
$$\left. * \left[\{(3K * \omega_{10}/\omega_{12}\}^{(1*\omega_{10})} \right] * \left[\{((1/3) * \omega_{10}/\omega_{13})\}^{(-1*\omega_{10})} \right] \right]^{1}. \tag{19.15}$$

The relationships between the primal and dual variables can be written as:

$$(\pi/4)KD^2H = \omega_{01}V \tag{19.16}$$

$$(\pi/12)KD^3 = \omega_{02}V \tag{19.17}$$

$$(4K)D^{-1} = \omega_{11}/\omega_{10} \tag{19.18}$$

$$(3K)H^{-1} = \omega_{12}/\omega_{10} \tag{19.19}$$

$$(1/3)DH^{-1} = \omega_{13}/\omega_{10}. \tag{19.20}$$

If one takes Equation (19.16) and divides by Equation (19.17), one obtains:

$$3H/D = \omega_{01}/\omega_{02}. \tag{19.21}$$

If one takes the inverse of Equation (19.20), one obtains:

$$3H/D = \omega_{10}/\omega_{13}. \tag{19.22}$$

Now comparing Equations (19.21) and (19.22), one can obtain an equation between the dual variables as:

$$\omega_{01}/\omega_{02} = \omega_{10}/\omega_{13}. \tag{19.23}$$

If one takes Equation (19.18) and divides by Equation (19.19), one obtains:

$$(4/3)H/D = \omega_{11}/\omega_{12}. \tag{19.24}$$

Now comparing Equations (19.21) and (19.24), one can obtain an additional equation between the dual variables as:

$$\omega_{01}/\omega_{02} = (9/4)\omega_{11}/\omega_{12}. \tag{19.25}$$

Now there are six equations with only the six dual variables and they are Equations (19.9), (19.10), (19.11), (19.13), (19.23), and (19.25). The procedure used was to solve for all of the variables in terms of ω_{02} and then obtain the specific value of ω_{02}.

From Equation (19.9), one obtains:

$$\omega_{01} = 1 - \omega_{02}. \tag{19.26}$$

If one adds Equations (19.10) and (19.11) one obtains:

$$3\omega_{01} + 3\omega_{02} - \omega_{11} - \omega_{12} = 0.$$

Which can be reduced to:

$$\omega_{11} + \omega_{12} = 3. \tag{19.27}$$

Using Equations (19.13) and (19.27) one obtains:

$$\omega_{10} = \omega_{11} + \omega_{12} - \omega_{13}$$

$$\omega_{10} = 3 - \omega_{13}. \tag{19.28}$$

Now using Equations (19.28) and (19.23) one obtains:

$$\omega_{01}/\omega_{02} = \omega_{10}/\omega_{13} = 3 - \omega_{13}/\omega_{13} = (1 - \omega_{02})/\omega_{02}.$$

Solving for ω_{13} one obtains:

$$\omega_{13} = 3\omega_{02}. \qquad (19.29)$$

From Equations (19.28) and (19.29):

$$\omega_{10} = 3 - \omega_{13} = 3 - 3\omega_{02}$$
$$\omega_{10} = 3(1 - \omega_{02}). \qquad (19.30)$$

Using Equations (19.26) and (19.29) in Equation (19.11), one obtains:

$$\omega_{12} = \omega_{01} + \omega_{13}$$
$$= (1 - \omega_{02}) + 3\omega_{02}$$
$$= 1 + 2\omega_{02}. \qquad (19.31)$$

Using Equations (19.10), (19.26), and (19.29) one has:

$$\omega_{11} = 2\omega_{01} + 3\omega_{02} - \omega_{13}$$
$$= 2(1 - \omega_{02}) + 3\omega_{02} - 3\omega_{02}$$
$$= 2(1 - \omega_{02}). \qquad (19.32)$$

Now using Equation (19.25) one can solve for ω_{02} using the values for ω_{01}, ω_{11}, and ω_{12}.

$$\omega_{01}/\omega_{02} = (9/4)\omega_{11}/\omega_{12}$$
$$(1 - \omega_{02})/\omega_{02} = (9/4)2(1 - \omega_{02})/(1 + 2\omega_{02}).$$

And solving for ω_{02} results in

$$\omega_{02} = 0.4.$$

Therefore,

$$\omega_{01} = 0.6$$
$$\omega_{11} = 1.2$$
$$\omega_{12} = 1.8$$
$$\omega_{13} = 1.2$$
$$\omega_{10} = 1.8.$$

The dual variables of the objective function of 0.6 and 0.4 indicate that the cylindrical part of the riser will be 60% of the volume and the hemispheric part will be 40% of the riser volume;

this could also be stated that the cylindrical part will be 50% more volume than the hemispheric part.

Now using the dual variables in Equation (19.15) to find the minimum volume one obtains:

$$
\begin{aligned}
V &= 1\left[\left[\{(\pi/4 * 1/\omega_{01})\}^{(1*\omega_{01})}\right] * \left[\{(\pi/12 * \omega_{02})\}^{(1*\omega_{02})}\right]\right.\\
&\quad * \left[\{(4K * \omega_{10}/\omega_{11})\}^{(1*\omega_{11})}\right] * \left[\{(3K * \omega_{10}/\omega_{12})\}^{(1*\omega_{12})}\right]\\
&\quad \left. * \left[\{((1/3) * \omega_{10}/\omega_{13})\}^{(-1*\omega_{13})}\right]\right]^{1}\\
&= 1\left[\left[\{(\pi/4) * 1/0.6)\}^{(1*0.6)}\right] * \left[\{(\pi/12 * 0.4)\}^{(1*0.4)}\right] * \left[\{(4K * 1.8/1.2)\}^{(1*1.2)}\right]\right.\\
&\quad \left. * \left[\{(3K * 1.8/1.8)\}^{(1*1.8)}\right] * \left[\{((1/3) * 1.8/1.2)\}^{(-1*1.2)}\right]\right]^{1}\\
&= 1\left[\left[\{((5/12)\pi)\}^{(0.6)}\right] * \left[\{((5/24)\pi)\}^{(0.4)}\right]\right.\\
&\quad \left. * \left[\{(6K)\}^{(1.2)}\right] * \left[\{(3K)\}^{(1.8)}\right] * \left[\{(1/2)\}^{(-1.2)}\right]\right]^{1}\\
&= 1\left[\left[\{(2 * 5/24)\pi)\}^{(0.6)}\right] * \left[\{((5/24)\pi)\}^{(0.4)}\right]\right.\\
&\quad \left. * \left[\{(2 * 3K)\}^{(1.2)}\right] * \left[\{(3K)\}^{(1.8)}\right] * \left[\{(2)\}^{(1.2)}\right]\right]^{1}\\
&= 2^{0.6}[(5/24)\pi]^{(0.6+0.4)} * 2^{1.2} * (3K)(1.2 + 1.8)2^{1.2}\\
&= (5\pi/24)^{1} * 2^{(0.6+1.2+1.2)} * (3K)^{3}\\
V &= (5\pi/24) * (6K)^{3}.
\end{aligned}
$$

(19.33)

The values for H and D can be found from Equations (19.18) and (19.19)

$$
\begin{aligned}
D &= 4K * (\omega_{10}/\omega_{11})\\
&= 4K * (1.8/1.2)\\
D &= 6K
\end{aligned}
$$

(19.34)

and

$$
\begin{aligned}
H &= 3K * (\omega_{10}/\omega_{12})\\
&= 3K * (1.8/1.8)\\
H &= 3K.
\end{aligned}
$$

(19.35)

Now the primal can be evaluated using Equation (19.7) with the values for H and D from Equations (19.34) and (19.35).

$$V = \pi D^2 H/4 + \pi D^3/12$$

$$V = \pi(6K)^2(3K)/4 + \pi(6K)^3/12$$

$$= \pi 27K^3 + \pi 18K^3 \tag{19.36}$$

$$= 45\pi K^3 \tag{19.37}$$

$$= (5\pi/24) * (6K)^3. \tag{19.38}$$

Note that the two terms for the volume in Equation (19.36) indicate that the volume of the cylindrical part is 50% more than the volume of the hemispherical part which is what was indicated by the dual variables. The values for the primal and dual are equivalent, which is required for solution validation. A better expression can be obtained as $6K$ is the value for the diameter for the cylindrical section where K is the modulus of the casing. This results in

$$V = (5\pi/24) * (6K)^3$$

$$= (5\pi/24)D^3. \tag{19.39}$$

This example illustrates that it is possible to solve problems with two degrees of difficulty in some instances, but there are numerous mathematical operations that must be performed to obtain the additional equations and solution.

19.3 DIMENSIONAL ANALYSIS TECHNIQUE FOR ADDITIONAL TWO EQUATIONS

Although it was relatively easy to determine the two additional equations, they also could have been obtained by dimensional analysis. The dimensional analysis approach sets up the primal dual relations of Equations (19.16)–19.20 and setting the primal variables on one side and the dual variables, constants, and objective function on the other side and giving the terms variable exponents as illustrated in Equation (19.40)

$$(D^2H)^A(D^3)^B(D^{-1})^C(H^{-1})^D(DH^{-1})^E = 1$$

$$= (\omega_{01}V/C_{01})^A(\omega_{02}V/C_{02})^B(\omega_{11}/\omega_{10}C_{11})^C(\omega_{12}/\omega_{10}C_{12})^D(\omega_{13}/\omega_{10}C_{13})^E, \tag{19.40}$$

where

$$C_{01} = (K\pi/4) \tag{19.41}$$

$$C_{02} = (K\pi/12) \tag{19.42}$$

$$C_{11} = 4K \tag{19.43}$$

$$C_{12} = 3K \tag{19.44}$$

$$C_{13} = 1/3. \tag{19.45}$$

One must balance the exponents to remove the primal variables (DH) and the dual objective function (V). An equation will also be written to consider ω_{10}. This is done by:

$$V \text{ values}\quad A + B \qquad\qquad\qquad = 0 \tag{19.46}$$

$$\omega_{10} \text{ values}\qquad\quad +C + D + E = 0 \tag{19.47}$$

$$D \text{ values}\quad 2A + 3B - C \quad\ \ + E = 0 \tag{19.48}$$

$$H \text{ values}\quad A \qquad\qquad - D - E = 0. \tag{19.49}$$

There are five variables and four equations, so one will find three variables in terms of the fourth variable. From Equation (19.46), one notes that

$$A = -B. \tag{19.50}$$

If Equations (19.48) and (19.49) are added and using the relation of (19.50), one obtains

$$C = -D. \tag{19.51}$$

Using the relationship of (19.51) and (19.47), one notes that

$$E = 0. \tag{19.52}$$

If one takes Equation (19.48) and subtracts two times Equation (19.46) and using $E = 0$, one obtains that

$$C = B. \tag{19.53}$$

Thus, if one lets $B = 1$, then $C = 1$, $D = -1$, $A = -1$, and $E = 0$. Using these values in (19.40) one obtains that

$$(\omega_{02}/\omega_{01})(\omega_{11}/\omega_{12}) = (C_{02}/C_{01})(C_{11}/C_{12})$$
$$= ((K\pi/12)/(K\pi/4))(4K)/(3K) = 4/9. \tag{19.54}$$

Upon rearranging terms, this is equivalent to Equation (19.25).

The second equation is obtained by allowing ω_{10} to be part of the new equation. Thus, take the Equations (19.46)–(19.49) and allow ω_{10} to enter the solution by making the $RHS = 1$ in Equation (19.47), which is renumbered as (19.55).

$$V \text{ values}\quad A + B \qquad\qquad\qquad = 0 \tag{19.46}$$

$$\omega_{10} \text{ values}\qquad\quad +C + D + E = 1 \tag{19.55}$$

$$D \text{ values}\quad 2A + 3B - C \quad\ \ + E = 0 \tag{19.48}$$

$$H \text{ values}\quad A \qquad\qquad - D - E = 0. \tag{19.50}$$

There are five variables and four equations, so one will find three variables in terms of the fourth variable. From Equation (19.46), one notes that

$$A = -B. \tag{19.56}$$

If Equations (19.48) and (19.49) are added and using the relation of (19.56), one obtains

$$C = -D. \tag{19.57}$$

Using the relationship of (19.55) and (19.57), one notes that

$$E = 1. \tag{19.58}$$

Using Equations (19.48) and (19.56), one obtains that

$$A = D + 1, \tag{19.59}$$

and thus

$$B = -(D + 1). \tag{19.60}$$

Using Equation (19.48) and (19.56), one obtains that

$$C = B + E = -D. \tag{19.61}$$

Thus, if one lets $D = 1$, then $A = 2$, $B = -2$, $C = -1$, and $E = 1$. Using these values in (19.40) one obtains that

$$(\omega_{01}V/C_{01})^2(\omega_{02}V/C_{02})^{-2}(\omega_{11}/\omega_{10}C_{11})^{-1}(\omega_{12}/\omega_{10}C_{12})^1(\omega_{13}/\omega_{10}C_{13})^1 = 1.$$

Reducing terms and putting the constants on the right hand side

$$(\omega_{01}/\omega_{02})^2(\omega_{12}/\omega_{11})(\omega_{13}/\omega_{10}) = (C_{12}C_{13}C_{01}^2)/(C_{02}^2C_{11})$$
$$= (3K * 1/3 * (K\pi/4)^2/((K\pi/12)^24K)$$
$$(\omega_{01}/\omega_{02})^2(\omega_{12}/\omega_{11})(\omega_{13}/\omega_{10}) = 9/4. \tag{19.62}$$

But rearranging Equation (19.25), one observes that

$$(\omega_{01}/\omega_{02})(\omega_{12}/\omega_{11}) = 9/4.$$

Cancelling these values from both sides, the result is:

$$(\omega_{01}/\omega_{02})(\omega_{13}/\omega_{10}) = 1, \tag{19.63}$$

which is equivalent to Equation (19.23).

$$\omega_{01}/\omega_{02} = \omega_{10}/\omega_{13}. \tag{19.23}$$

Thus, the dimensional analysis approach gives the same results as the substitution approach for obtaining the additional equations. The advantage of the dimensional analysis approach is that it is more straightforward in obtaining the relationships.

The design parameters determined are for the case where solidification time is the controlling constraint. In many cases feed metal is the controlling constraint and this would require another solution.

19.4   EVALUATIVE QUESTIONS

1. A side riser with a hemispheric top is to be designed for a rectangular casting (5 cm * 8 cm * 10 cm) which has a surface area of 340 cm² and a volume of 400 cm³. The hot metal cost is 100 Rupees per kg and the metal density is 3.0 gm/cm³. Compare these results with Problem 1 in Section 8.4.

   a) What are the dimensions of the hemispherical side riser (H and D)?

   b) What is the volume of the hemispherical side riser (cm³)?

   c) What is the metal cost of the hemispherical side riser (Rupees)?

   d) What is the metal cost of the casting (Rupees)?

2. Two castings of equal volume but of different dimensions are to be cast. If one is a 3 in cube and the other is a plate of $1 \times 3 \times 9$ inches and a top riser is to be used, what are the dimensions (H and D) of the risers for the two cases?

19.5   REFERENCES

[1] R. C. Creese, Optimal riser design by geometric programming, *AFS Cast Metals Research Journal*, Vol. 7, pp. 118–121, 1971.

[2] R. C. Creese, Dimensioning of risers for long freezing range alloys by geometric programming, *AFS Cast Metals Research Journal*, Vol. 7, pp. 182–184, 1971.

[3] R. C. Creese, Generalized riser design by geometric programming, *AFS Transactions*, Vol. 87, pp. 661–664, 1979.

[4] R. C. Creese, An evaluation of cylindrical riser designs with insulating materials, *AFS Transactions*, Vol. 87, pp. 665–669, 1979.

[5] R. C. Creese, Cylindrical top riser design relationships for evaluating insulating materials, *AFS Transactions*, Vol. 89, pp. 345–348, 1981.

CHAPTER 20

Liquefied Petroleum Gas (LPG) Cylinders Case Study

20.1 INTRODUCTION

This case study problem deals with the design of liquefied petroleum gas cylinders, more commonly known as propane gas cylinders in the U.S. This is a very interesting problem as it has two general solutions as well as one degree of difficulty. The two general solutions occur depending upon the relationship between the constants. What happens in this particular problem is that one of the two constraints can be either binding or loose depending upon the value of the constants. This problem was initially presented by Ravivarna Pericherla [1] in his M.S. thesis and later published with his advisor Dr. Gopalakrishnan [2] where geometric programming was used only for the part of the solution which had zero degrees of difficulty. This case study utilizes geometric programming for both solutions.

20.2 PROBLEM FORMULATION

The problem was to minimize the drawing force (Z) to produce the tank by deep drawing, and two constraints were considered so the tank would have a minimum volume and the height/diameter ratio would be less than one. The formulation of the primal problem was:

$$\text{Minimize } Z = K_1 hd + K_2 d^2, \tag{20.1}$$

subject to the two constraints:

$$\pi d^2 h/4 \geq V_{\min}, \tag{20.2}$$

or in the proper geometric programming form as

$$(4V_{\min}/\pi)d^{-2}h^{-1} \leq 1, \tag{20.3}$$

and

$$h/d \leq 1, \tag{20.4}$$

where

$$K_1 = \pi PYC/F, \tag{20.5}$$

and

$$K_2 = (C - E)\pi PY/2F, \tag{20.6}$$

where

Z = drawing force

P = internal gas pressure

Y = material yield strength

F = hoop stress

C = constant = 1.04

E = constant = 0.65.

If all the constants are combined, the primal form can be written as:

$$Z = K_1 hd + K_2 d^2, \tag{20.1}$$

subject to:

$$K_3 h^{-1} d^{-2} \leq 1, \tag{20.7}$$

and

$$K_4 hd^{-1} \leq 1, \tag{20.8}$$

where

$$K_3 = (4V_{\min}/\pi), \tag{20.9}$$

and

$$K_4 = 1 \text{ (This could be taken as the minimum } d/h \text{ ratio).} \tag{20.10}$$

From the coefficients and signs, the signum values for the dual are:

$$\sigma_{01} = 1$$
$$\sigma_{02} = 1$$
$$\sigma_{11} = 1$$
$$\sigma_{21} = 1$$
$$\sigma_1 = 1$$
$$\sigma_2 = 1.$$

The dual problem formulation is:

$$\text{Objective Function} \qquad \omega_{01} \;+ \omega_{02} \qquad\qquad\qquad = 1 \qquad\qquad (20.11)$$

$$h \text{ terms} \qquad\qquad\qquad \omega_{01} \qquad\quad - \;\omega_{11} + \omega_{21} = 0 \qquad\qquad (20.12)$$

$$d \text{ terms} \qquad\qquad\qquad \omega_{01} + 2\omega_{02} - 2\omega_{11} - \omega_{21} = 0. \qquad\qquad (20.13)$$

The degrees of difficulty are equal to:

$$D = T - (N + 1) = 4 - (2 + 1) = 1. \qquad\qquad (20.14)$$

From the constraint equations which have only one term it is apparent that:

$$\omega_{10} = \omega_{11} \qquad\qquad (20.15)$$

and

$$\omega_{20} = \omega_{21}. \qquad\qquad (20.16)$$

An additional equation is needed, and one must examine the primal dual relationships to find the additional relationship, which are:

$$K_1 h d = \omega_{01} Z \qquad\qquad (20.17)$$

$$K_2 d^2 = \omega_{02} Z \qquad\qquad (20.18)$$

$$K_3 h^{-1} d^{-2} = \omega_{11}/\omega_{10} = 1 \qquad\qquad (20.19)$$

$$K_4 h d^{-1} = \omega_{21}/\omega_{20} = 1. \qquad\qquad (20.20)$$

If one adds Equations (20.12) and (20.13), one can solve for ω_{11} by:

$$2\omega_{01} + 2\omega_{02} - 3\omega_{11} = 0 \qquad\qquad (20.21)$$

$$2(\omega_{01} + \omega_{02}) - 3\omega_{11} = 0 \qquad\qquad (20.22)$$

$$2 - 3\omega_{11} = 0 \qquad\qquad (20.23)$$

$$\omega_{11} = 2/3. \qquad\qquad (20.24)$$

If one takes Equation (20.17) and divides it by Equations (20.18) and (20.20), one obtains:

$$K_1 h d / \left(K_2 d^2 * K_4 h d^{-1} \right) = K_1/(K_2 K_4) = \omega_{01} Z/(\omega_{02} Z * 1) = \omega_{01}/\omega_{02}. \qquad (20.25)$$

This can be solved for ω_{01} in terms of ω_{02} and the constants:

$$\omega_{01} = \omega_{02}(K_1/(K_2 K_4)). \qquad\qquad (20.26)$$

Now using Equations (20.26) and (20.11) one can solve for ω_{01} and ω_{02} and obtain:

$$\omega_{01} = K_1/(K_1 + K_2 K_4) \qquad\qquad (20.27)$$

$$\omega_{02} = K_2 K_4/(K_1 + K_2 K_4). \qquad\qquad (20.28)$$

Now using Equation (20.12) and substituting the values for ω_{01} and ω_{11} one obtains:

$$\omega_{21} = \omega_{11} - \omega_{01} = (2K_2 K_4 - K_1)/[3(K_1 + K_2 K_4)]. \tag{20.29}$$

Now ω_{21} must be ≥ 0 so that implies that

$$2K_2 K_4 - K_1 \geq 0, \tag{20.30}$$

or

$$K_2 K_4 / K_1 \geq 1/2. \tag{20.31}$$

Thus, there are two sets of solutions depending upon whether Equation (20.31) holds; this can be illustrated in Figure 20.1.

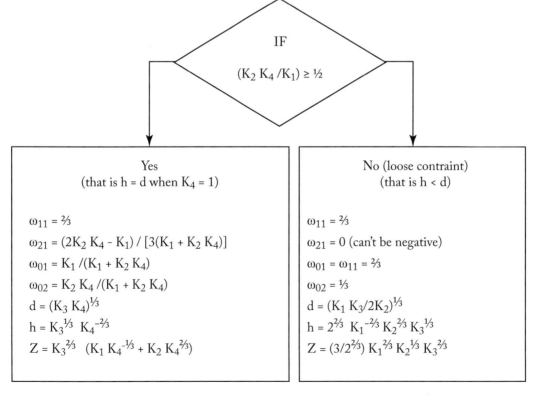

Figure 20.1: Dual and primal design equations for the two solutions.

Now the solutions can be found for the two cases. If the answer is "No," then the objective function can be evaluated using the dual expression:

$$Z = d(\omega) = \sigma \left[\prod_{m=0}^{M} \prod_{t=1}^{T_m} (C_{mt}\omega_{mo}/\omega_{mt})^{\sigma_{mt}\omega_{mt}} \right]^{\sigma}, \tag{20.32}$$

where by definition

$$\omega_{00} = 1 \tag{20.33}$$

$$Z = [K_1 * 1/(2/3)]^{1*2/3}[K_2 * 1/(1/3)]^{1*1/3}[K_3 * (2/3)/(2/3)]^{1*2/3}, \tag{20.34}$$

which can be reduced to:

$$Z = \left[3/2^{2/3} \right] K_1^{2/3} K_2^{1/3} K_3^{2/3}. \tag{20.35}$$

Now the primal variables can be determined from the primal dual relationships. If one uses Equation (20.18) and solves for d^2 and then for d, one obtains:

$$d^2 = \omega_{02}Z/K_2 = (1/3) * \left\{ \left[3/2^{2/3} \right] K_1^{2/3} K_2^{1/3} K_3^{2/3} \right\} / K_2 = (K_1 K_3/2K_2)^{2/3}. \tag{20.36}$$

Solving for d results in:

$$d = (K_1 K_3/2K_2)^{1/3}. \tag{20.37}$$

Similarly, if one uses Equation (20.17) to solve for h one obtains:

$$h = \omega_{01} Z/K_1 d$$
$$= \left\{ (2/3) * \left\{ \left[3/2^{2/3} \right] K_1^{2/3} K_2^{1/3} K_3^{2/3} \right\} \right\} / \left[K_1 * (K_1 K_3/2K_2)^{1/3} \right]. \tag{20.38}$$

Reducing terms one obtains:

$$h = 2^{2/3} K_1^{-2/3} K_2^{2/3} K_3^{1/3}. \tag{20.39}$$

Now substituting the primal variables into the primal objective function, one has:

$$Z = K_1 hd + K_2 d^2 \tag{20.1}$$
$$= K_1 * 2^{2/3} K_1^{-2/3} K_2^{2/3} K_3^{1/3} * (K_1 K_3/2K_2)^{1/3} + K_2 * (K_1 K_3/2K_2)^{2/3}$$
$$= 2K_1^{2/3} K_2^{1/3} K_3^{2/3}/2^{2/3} + K_1^{2/3} K_2^{1/3} K_3^{2/3}/2^{2/3}$$
$$= 3K_1^{2/3} K_2^{1/3} K_3^{2/3}/2^{2/3}$$
$$= \left[3/2^{2/3} \right] * K_1^{2/3} K_2^{1/3} K_3^{2/3}. \tag{20.40}$$

The equations for the primal and dual, Equations (20.35) and (20.40), give the same results. Thus, one has a general solution for the objective function and the two primal variables when the

"No" route was taken. The dual variables indicate that the first term cost will always be twice the magnitude of the second term.

The "Yes" route has the dual variables as functions of the constants and thus is the more complex route. The dual variables are a function of the constants and the cost ratio between the first and second terms will not be constant but vary as $K_1 : (K_2 K_4)$. The objective function can be evaluated using the dual expression:

$$Z = d(\omega) = \sigma \left[\prod_{m=0}^{M} \prod_{t=1}^{T_m} (C_{mt} \omega_{mo} / \omega_{mt})^{\sigma_{mt} \omega_{mt}} \right]^{\sigma} \tag{20.32}$$

where by definition

$$\omega_{00} = 1. \tag{20.33}$$

The dual variables are:

$$\omega_{11} = 2/3 \tag{20.41}$$
$$\omega_{21} = (2 K_2 K_4 - K_1)/[3(K_1 + K_2 K_4)] \tag{20.42}$$
$$\omega_{01} = K_1/(K_1 + K_2 K_4) \tag{20.43}$$
$$\omega_{02} = K_2 K_4/(K_1 + K_2 K_4). \tag{20.44}$$

Using these dual variables and the signum values the dual objective function is:

$$\begin{aligned}
Z &= (K_1 \omega_{00}/\omega_{01})^{1*\omega_{01}} (K_2 \omega_{00}/\omega_{02})^{1*\omega_{02}} (K_3)^{1*\omega_{11}} (K_4)^{1*\omega_{21}} \\
&= (K_1 + K_2 K_4)^{\omega_{01}} [(K_1 + K_2 K_4)/K_4]^{\omega_{02}} (K_3)^{2/3} (K_4)^{\omega_{11} - \omega_{01}} \\
&= [(K_1 + K_2 K_4)/K_4]^{\omega_{01}} [(K_1 + K_2 K_4)/K_4]^{\omega_{02}} (K_3 K_4)^{2/3} \\
&= [(K_1 + K_2 K_4)/K_4](K_3 K_4)^{2/3} \\
&= (K_3)^{2/3} \left[\left(K_1 K_4^{-1/3} + K_2 K_4^{2/3} \right) \right].
\end{aligned} \tag{20.45}$$

The primal variables can be obtained from the relationships between the primal and dual variables. Using Equation (20.20) one can specify a relation between h and d via K_4 which is:

$$h = d/K_4. \tag{20.46}$$

If one combines Equations (20.19) and (20.20) one obtains:

$$\left(K_3 h^{-1} d^{-2} \right) \left(K_4 h d^{-1} \right) = 1 * 1$$

which yields:

$$K_3 K_4 d^{-3} = 1$$

or

$$d = (K_3 K_4)^{1/3}. \tag{20.47}$$

Now from Equations (20.46) and (20.47) one obtains:

$$h = d/K_4 = K_3^{1/3} K_4^{-2/3}. \tag{20.48}$$

Now using the primal equation for the objective function with the primal variables one has:

$$\begin{aligned}
Z &= K_1 h d + K_2 d^2 \tag{20.1} \\
&= K_1 \left(K_3^{1/3} K_4^{-2/3} \right) (K_3 K_4)^{1/3} + K_2 (K_3 K_4)^{2/3} \\
&= K_1 \left(K_3^{2/3} K_4^{-1/3} \right) + K_2 (K_3 K_4)^{2/3} \\
&= K_3^{2/3} \left(K_1 K_4^{-1/3} + K_2 K_4^{2/3} \right). \tag{20.49}
\end{aligned}$$

Note that the general objective function is the same for both the primal and dual solutions. This problem illustrates that it is possible to solve a problem with one degree of difficulty and have two solutions depending upon the specific values of the constants in the problem.

20.3 DIMENSIONAL ANALYSIS TECHNIQUE FOR ADDITIONAL EQUATION

It was not very difficult to determine the additional equation by repeated substitution, but the technique of dimensional analysis can be used to verify it and this will be demonstrated to obtain Equation (20.26). The dimensional analysis approach sets up the primal dual relations of Equations (20.17)–(20.20) and setting the primal variables on one side and the dual variables, constants, and objective function on the other side and giving the terms variable exponents as illustrated in Equation (20.50).

$$(hd)^A (d^2)^B (h^{-1} d^{-2})^C (hd^{-1})^D = 1 = (\omega_{01} Z/K_1)^A (\omega_{02} Z/K_2)^B (1/K_3)^C (1/K_4)^D. \tag{20.50}$$

One must balance the exponents to remove the primal variables (d, h) and the dual objective function (Z). This is done by:

Z values	$A + B = 0$	(20.51)
h values	$A \quad - C + D = 0$	(20.52)
d values	$A + 2B - 2C - D = 0.$	(20.53)

There are four variables and three equations, so one will find three variables in terms of the fourth variable. From Equation (20.51), one obtains

$$A = -B. \tag{20.54}$$

Using Equations (20.54) and (20.52), one notes that

$$B = D - C. \tag{20.55}$$

If one subtracts Equation (20.52) from Equation (20.53), one obtains that

$$B = D + C/2. \tag{20.56}$$

If one compares Equations (20.55) and (20.56), the only solution for both equations to be satisfied is if:

$$C = 0. \tag{20.57}$$

If one sets $D = 1$, then $B = 1$ and $A = -1$. Using these values in Equation (20.50) one obtains

$$(hd)^{-1}(d^2)^1(h^{-1}d^{-2})^0(hd^{-1})^1 = 1 = (\omega_{01}Z/K_1)^{-1}(\omega_{02}Z/K_2)^1(1/K_3)^0(1/K_4)^1. \tag{20.58}$$

This results in Equation (20.59) which is equivalent to Equation (20.26)

$$\omega_{01} = \omega_{02}(K_1/(K_2K_4)). \tag{20.59}$$

The solution to the problem is the same, but the obtaining of the additional equation is often the difficult step in the process. The dimensional analysis technique is often easier than the substitution method as it is more consistent and often faster in obtaining the necessary additional equation(s).

20.4 EVALUATIVE QUESTIONS

1. A tank is to be designed with a minimum volume of 17,500,000 mm^3 and the values for parameters are:

$$P = 0.2535 \text{ kg/mm}^2$$
$$F = 32.33 \text{ kg/mm}^2$$
$$Y = 25 \text{ kg/mm}^2$$
$$C = 1.04$$
$$E = 0.65.$$

Determine the amount of the drawing force (kg) and the height (mm) and diameter (mm) of the tank.

2. A new procedure was developed for another older machine which changed the expression for K_2. The new expression was:

$$K_2 = (2C - E)\pi PY/(2F). \tag{20.60}$$

Using the same data as in Evaluative Question 1, determine the amount of the drawing force (kg) and the height (mm) and diameter (mm) of the new tank.

3. A tank is to be designed with a minimum volume of 11,000 in³ and the values for parameters are:

$$P = 360 \text{ lb/in}^2$$
$$F = 46,000 \text{ lb/in}^2$$
$$Y = 35,500 \text{ lb/in}^2$$
$$C = 1.04$$
$$E = 0.65.$$

Determine the amount of the drawing force (lb and tons) and the height (in) and diameter (in) of the tank.

20.5 REFERENCES

[1] R. Pericherla, *Design and manufacture of liquefied petroleum gas cylinders*, M.S. thesis, West Virginia University, Morgantown, WV, 1992. 131

[2] B. Gopalakrishnan and R. Pericherla, Design for manufacturability of the liquefied petroleum gas cylinder, *Proc. of the International Industrial Engineering Research Conference*, IIE, Miami, FL, 1997. 131

CHAPTER 21

Material Removal/Metal Cutting Economics with Two Constraints Case Study

21.1 INTRODUCTION

The material removal metal cutting economics case study in Chapter 12 had only a feed constraint, but the problem presented here also includes the horsepower constraint. The additional constraint leads to a problem with one degree of difficulty and to the possibility of multiple solutions similar to that in the liquefied petroleum gas cylinder case study. Thus, the problem is more difficult to solve and needs to be considered separately, but the equations to define the problem are similar to those of Chapter 12. The theory is from the paper by Tsai [1] and the example data are from that of Ermer [2].

21.2 PROBLEM FORMULATION

The problem is based upon the modified form of the Taylor's Tool Life Equation and two constraints, the feed rate and horsepower constraints. The cost function represents the sum of the various costs and is expressed as:

$$C_u = (R_o + R_m)t_h + (R_o + R_m)t_c + C_t n_t + (R_o + R_m)t_{ch}n_t, \tag{21.1}$$

where

C_u = Unit Cost, \$/piece

R_o = Operator Rate, \$/min

R_m = Machine Rate, \$/min

C_t = Tool Cost (per cutting edge for insert tools), \$

t_h = Handling Time (to insert and remove piece), min/piece

t_c = Cutting Time, min/piece

t_{ch} = Tool Changing Time (to change tool), min

n_t = Number of tool changes per piece.

The modified form of Taylor's Tool Life Equation is:

$$TV^{1/n}f^{1/m} = C, \tag{21.2}$$

where

T = Tool Life, min

V = Cutting Speed, feet/min

$1/n$ = Cutting Speed Exponent

f = Feed Rate, in/rev

$1/m$ = Feed Rate Exponent

C = Taylor's Modified Tool Life Constant (min).

The cutting time can be expressed as:

$$t_c = Bf^{-1}V^{-1}, \tag{21.3}$$

where

B = Cutting Path Surface Factor (an input value), ft

f = Feed Rate, in/rev

V = Cutting Speed, feet/min.

The number of tool changes per piece can be found by:

$$n_t = Qt_c/T, \tag{21.4}$$

where

Q = Fraction of cutting time that the tool is worn

t_c = Cutting Time, min/piece

T = Tool Life, min.

The value of Q is approximately 1.0 for operations such as turning, but for other material removal/metal cutting operations it may be as low as 0.10, say for horizontal milling operations. Utilizing Equations (21.2) and (21.3) in Equation (21.4), one obtains:

$$n_t = QBC^{-1}f^{(1/m-1)}V^{(1/n-1)}. \tag{21.5}$$

The total cost can be expressed as the sum of the cost components as:

$$C_u = \text{Machine Cost} + \text{Operator Cost} + \text{Tool Cost} + \text{Tool Changing Cost}$$
$$= R_m(t_c + t_l) + R_o(t_c + t_l) + C_t n_t + (R_o + R_m)t_{ch}n_t. \tag{21.6}$$

After substituting for n_t, t_c, and T to obtain the cost expression in terms of V and f, one obtains:

$$C_u = (R_o + R_m)t_l + (R_o + R_m)Bf^{-1}V^{-1}$$
$$+ [(R_o + R_m)t_{ch} + C_t]QBC^{-1} f^{(1/m-1)}V^{(1/n-1)}. \tag{21.7}$$

The constants can be combined and the unit cost objective function can be restated as:

$$C_u = K_{oo} + K_{01}f^{-1}V^{-1} + K_{02}f^{(1/m-1)}V^{(1/n-1)}, \tag{21.8}$$

where

$$K_{oo} = (R_o + R_m)t_l$$
$$K_{01} = (R_o + R_m)B$$
$$K_{02} = [(R_o + R_m)t_{ch} + C_t]QBC^{-1}.$$

Since K_{oo} is a constant, the objective function to be minimized is:

$$Y = K_{01}f^{-1}V^{-1} + K_{02}f^{(1/m-1)}V^{(1/n-1)}, \tag{21.9}$$

where

$$Y = \text{Variable portion of the unit cost.}$$

The two constraints must be developed into geometric programming format. The feed constraint is:

$$f \leq f_{max}, \tag{21.10}$$

where

$$f_{max} = \text{Maximum Feed Limit (in/rev).}$$

The feed constraint is typically used to control the surface finish as the smaller the feed, the better the surface finish.

The general form of the horsepower constraint is given as:

$$aV^b f^c \leq Hp, \tag{21.11}$$

where

$a = $ Horsepower Constraint Constant

$b = $ Velocity Exponent for Horsepower Constraint

$c = $ Feed Rate Exponent for Horsepower Constraint

$HP = $ Horsepower Limit.

These constraints can be put into geometric programming form as:

$$K_{11} f \leq 1 \tag{21.12}$$

$$K_{21} V^b f^c \leq 1, \tag{21.13}$$

where

$$K_{11} = 1/f_{\max}$$
$$K_{21} = a/Hp.$$

21.3 PROBLEM SOLUTION

The primal problem can be stated as:

$$\text{Minimize:} \quad Y = K_{01} f^{-1} V^{-1} + K_{02} f^{(1/m-1)} V^{(1/n-1)}. \tag{21.14}$$

Subject to the constraints:

$$K_{11} f \leq 1 \tag{21.15}$$

$$K_{21} V^b f^c \leq 1. \tag{21.16}$$

One issue of concern is that the tool life equation and horsepower equations be compatible for the same material so that the exponents $n, m, b,$ and c apply to the same conditions. Often the equations were developed for different material-tool-machine conditions and the combined equations are not correct for the situation and give poor results.

From the coefficients and signs, the signum values for the dual are:

$$\sigma_{01} = 1$$
$$\sigma_{02} = 1$$
$$\sigma_{11} = 1$$
$$\sigma_{21} = 1$$
$$\sigma_{1} = 1$$
$$\sigma_{2} = 1$$

The dual problem can be formulated as:

$$\omega_{01} \qquad + \omega_{02} \qquad\qquad = 1 \tag{21.17}$$

$$f \text{ terms} \quad -\omega_{01} + (1/m - 1)\omega_{02} + \omega_{11} + c\omega_{21} = 0 \tag{21.18}$$

$$V \text{ terms} \quad -\omega_{01} + (1/n - 1)\omega_{02} \qquad + b\omega_{21} = 0. \tag{21.19}$$

The degrees of difficulty (D) are equal to:

$$D = T - (N + 1) = 4 - (2 + 1) = 1. \tag{21.20}$$

From the constraint equations which have only one term it is apparent that:

$$\omega_{10} = \omega_{11} \tag{21.21}$$

$$\omega_{20} = \omega_{21}. \tag{21.22}$$

The dual objective function is:

$$Y = (K_{01}/\omega_{01})^{\omega_{01}} (K_{02}/\omega_{02})^{\omega_{02}} (K_{11})^{\omega_{11}} (K_{21})^{\omega_{21}}. \tag{21.23}$$

Thus, we have four variables and three equations, or one degree of difficulty. To find an additional equation one must get the number of variables reduced to one in the dual objective function, and differentiate the logarithm of the equation to obtain another equation by setting the derivative to zero. This procedure is known as the "*constrained derivative*" approach. It was decided to determine the dual variables in terms of ω_{02}.

From Equation (21.17) one obtains:

$$\omega_{01} = 1 - \omega_{02}. \tag{21.24}$$

From Equation (21.19) one obtains:

$$\begin{aligned} b\omega_{21} &= \omega_{01} - (1/n - 1)\omega_{02} \\ &= (1 - \omega_{02}) - (1/n - 1)\omega_{02} \\ &= 1 - \omega_{02}/n \\ \omega_{21} &= 1/b - \omega_{02}/bn. \end{aligned} \tag{21.25}$$

From Equation (21.18) one obtains:

$$\begin{aligned} \omega_{11} &= \omega_{01} - (1/m - 1)\omega_{02} - c\omega_{21} \\ &= 1 - \omega_{02} - (1/m - 1)\omega_{02} - c(1/b - \omega_{02}/bn) \\ &= (1 - c/b) + \omega_{02}(c/bn - 1/m) \\ \omega_{11} &= (1 - c/b) + \omega_{02}Z, \end{aligned} \tag{21.26}$$

where

$$Z = (c/bn - 1/m). \tag{21.27}$$

The dual can be written as:

$$Y = (K_{01}/(1 - \omega_{02}))^{(1-\omega_{02})} (K_{02}/\omega_{02})^{\omega_{02}} (K_{11})^{((1-c/b)+\omega_{02}Z)} (K_{21})^{(1/b-\omega_{02}/bn)}. \tag{21.28}$$

The log of the dual is:

$$\begin{aligned} \text{Log}(Y) = {} &(1 - \omega_{02}) \log(K_{01}/(1 - \omega_{02})) + \omega_{02} \log(K_{02}/\omega_{02}) \\ &+ (1 - c/b) \log(K_{11}) + \omega_{02}Z \log(K_{11}) \\ &+ 1/b \log(K_{21}) - \omega_{02}/bn \log(K_{21}). \end{aligned} \tag{21.29}$$

Find $\partial(\log Y)/\partial\omega_{02}$ and set it to zero to find an additional equation.

$$\partial(\log Y)/\partial\omega_{02} = (1 - \omega_{02})[1/\{K_{01}/(1 - \omega_{02})\}]\left[K_{01}(-1)(1 - \omega_{02})^{-2}(-1)\right]$$
$$+ [\log(K_{01}/(1 - \omega_{02}))](-1)$$
$$+ \omega_{02}[1/\{K_{02}/\omega_{02}\}]\left[K_{02}(-1)\omega_{02}^{-2}(1)\right] + [\log(K_{02}/\omega_{02})](1)$$
$$+ 0 + Z\log(K_{11}) + 0 - (1/bn)\log(K_{21})$$

$$\partial(\log Y)/\partial\omega_{02} = (-1) - [\log(K_{01}/(1 - \omega_{02}))] + 1 + [\log(K_{02}/\omega_{02})]$$
$$+ 0 + Z\log(K_{11}) + 0 - (1/bn)\log(K_{21})$$

$$\partial(\log Y)/\partial\omega_{02} = \log[(K_{02}/K_{01})((1 - \omega_{02})/(\omega_{02}))] + \log(K_{11})^{Z} - \log(K_{21})^{1/bn}$$

$$\partial(\log Y)/\partial\omega_{02} = \log[(K_{02}/K_{01})((1 - \omega_{02})/(\omega_{02}))] + \log\left[(K_{11})^{Z}/\log(K_{21})^{1/bn}\right]$$

$$\partial(\log Y)/\partial\omega_{02} = \log\left[(K_{02}/K_{01})((1 - \omega_{02})/(\omega_{02}))(K_{11})^{Z}/(K_{21})^{1/bn}\right]. \tag{21.30}$$

Setting the expression equal to zero and then use the anti-log, one goes through various steps to obtain an expression for ω_{02}:

$$1 = (K_{02}/K_{01})((1 - \omega_{02})/(\omega_{02}))(K_{11})^{Z}/(K_{21})^{1/bn}$$

$$(1 - \omega_{02})/(\omega_{02}) = (K_{01}/K_{02})\left((K_{21})^{1/bn}/(K_{11})^{Z}\right)$$

$$1/\omega_{02} - 1 = (K_{01}/K_{02})\left((K_{21})^{1/bn}/(K_{11})^{Z}\right)$$

$$1/\omega_{02} = (K_{01}/K_{02})\left((K_{21})^{1/bn}/(K_{11})^{Z}\right) + 1$$

$$\omega_{02} = \left(K_{02}K_{11}^{Z}\right)/\left(K_{01}K_{21}^{1/bn} + K_{02}K_{11}^{Z}\right). \tag{21.31}$$

Now the values for other three dual variables can be obtained in terms of ω_{02} as:

$$\omega_{01} = 1 - \omega_{02} \tag{21.32}$$
$$\omega_{21} = 1/b - \omega_{02}/bn \tag{21.33}$$
$$\omega_{11} = (1 - c/b) + \omega_{02}Z. \tag{21.34}$$

Substituting the values of the dual variables into the dual objective function, the dual objective function can be expressed in terms of ω_{02} and is:

$$Y = (K_{01}/(1 - \omega_{02}))^{(1-\omega_{02})}(K_{02}/\omega_{02})^{\omega_{02}}(K_{11})^{((1-c/b)+\omega_{02}Z)}(K_{21})^{(1/b-\omega_{02}/bn)} \tag{21.35}$$

and the minimum cost would be:

$$C_u = Y + K_{00} \tag{21.36}$$

or

$$C_u = K_{00} + (K_{01}/(1 - \omega_{02}))^{(1-\omega_{02})}(K_{02}/\omega_{02})^{\omega_{02}}$$
$$(K_{11})^{((1-c/b)+\omega_{02}Z)}(K_{21})^{(1/b-\omega_{02}/bn)}. \tag{21.37}$$

By using the primal-dual relationships of the objective function, one can obtain the following equation for the primal variables. Starting with

$$K_{01} f^{-1} V^{-1} = \omega_{01} Y \tag{21.38}$$

and

$$K_{02} f^{1/m-1} V^{1/n-1} = \omega_{02} Y. \tag{21.39}$$

One can obtain equations for the primal variables in terms of the dual variables, dual objective function, and constants. These equations are for the case when both constraints are binding.

$$f = (\omega_{02} K_{01}/(\omega_{01} K_{02}))^m (K_{01}/\omega_{01})^{(m(m-1)/(n-m))}$$
$$(\omega_{02}/K_{02})^{((mm)/(n-m))} Y^{(m/(n-m))} \tag{21.40}$$
$$V = (K_{01}/\omega_{01})^{(n(m-1)/(m-n))} (\omega_{02}/K_{02})^{((mn)/(m-n))} Y^{(n/(m-n))}. \tag{21.41}$$

Upon examination of the equations for the dual variables, it is possible for either ω_{11} or ω_{21} to be negative, which means that that constraint would not be binding. Thus, the dual variable must be set to zero if the constraint is not binding. If ω_{11} is zero (feed constraint is not binding), then Equations (21.17), (21.18), and (21.19) become:

$$\omega_{01} + \quad\quad \omega_{02} \quad\quad = 1 \tag{21.42}$$
$$f \text{ terms} \quad -\omega_{01} + (1/m - 1)\omega_{02} + c\omega_{21} = 0 \tag{21.43}$$
$$V \text{ terms} \quad -\omega_{01} + (1/n - 1)\omega_{02} + b\omega_{21} = 0. \tag{21.44}$$

The degree of difficulty becomes zero with the variable ω_{11} removed, and the new values for the dual variables become:

$$\omega_{02} = (b - c)/(b/m - c/n) \tag{21.45}$$
$$\omega_{01} = 1 - \omega_{02} \tag{21.46}$$
$$\omega_{11} = 0 \text{ (loose constraint)} \tag{21.47}$$
$$\omega_{21} = (1/m - 1/n)/(b/m - c/n). \tag{21.48}$$

One must first check to determine if all the dual variables are positive as the equations may not be valid for the conditions. If ω_{02} is negative the exponents (b, c, m, and n) were probably not determined in the same study. The horsepower equation also restricts the feed as well as the cutting speed, so this model is rarely applied.

If ω_{21} is negative (horsepower constraint is not binding), then Equations (21.17), (21.18), and (21.19) become:

$$-\omega_{01} + \qquad\qquad \omega_{02} \qquad = 1 \qquad\qquad (21.49)$$

$$f \text{ terms} \qquad -\omega_{01} + (1/m - 1)\omega_{02} + \omega_{11} = 0 \qquad\qquad (21.50)$$

$$V \text{ terms} \qquad -\omega_{01} + (1/n - 1)\omega_{02} \qquad = 0. \qquad\qquad (21.51)$$

The degree of difficulty becomes zero with the variable ω_{21} removed, and the new values for the dual variables become:

$$\omega_{02} = n \qquad\qquad (21.52)$$

$$\omega_{01} = 1 - \omega_{02} = 1 - n \qquad\qquad (21.53)$$

$$\omega_{11} = 1 - n/m \qquad\qquad (21.54)$$

$$\omega_{21} = 0. \qquad\qquad (21.55)$$

The equations for velocity and feed for this set of dual variables are:

$$V = (n/(1-n))^n (K_{01}/K_{02})^n K_{11}^{(n/m)} \qquad\qquad (21.56)$$

$$f = (K_{01}/(\omega_{01} YV)). \qquad\qquad (21.57)$$

Thus, there are three different solutions based upon the values of the constants in the problem. However, the equations for the dual variables for the three possibilities have been obtained and the only question is to determine which of the three conditions are applicable. The results for the primal variables have been given for the case when both constraints are binding and when the feed constraint is binding (horsepower is non-binding). The horsepower constraint is often not binding, but the feed constraint is usually always a binding constraint. The dual variables are functions of the tool life equation exponents only when the horsepower constraint is not binding and the degrees of difficulty are zero. The dual variables are functions of both the exponents and constants when the degrees of difficulty are one.

21.4   EXAMPLE PROBLEM

An example problem will be used to illustrate the application of the formulas and the magnitude of the results obtained. The data used is from Ermer [2] and results obtained are in Table 21.1.

The model results indicate that the cutting speed is 290 ft/min and the feed rate is at the max of 0.005 in/rev. The variable cost (Y) is $1.05 and the total cost is $1.25.

Table 21.1: Data for metal cutting example problem

Input Parameters		Calculated Constants and Variables	
Q	1	K_{00}	0.20
R_0	0.04 \$/min	K_{01}	1.2566
R_m	0.06 \$/min	K_{02}	$1.80 * 10^{-8}$
C_t	0.50 \$/edge	K_{11}	200
t_h	2 min	K_{21}	0.358
t_{ch}	0.50 min	Z	2.2684
D	6 inches	ω_{01}	0.8216
L	8 inches	ω_{02}	0.1784
m	0.862	ω_{11}	0.5476
n	0.25	ω_{21}	0.3146
C	$3.84 * 10^8$ min	Y	\$ 1.05
a	3.58	C_u	\$ 1.25
b	0.91	B	12.56 in-ft = πD (in) L(ft)
c	0.78	f	0.005 in/rev
Hp	10	V	290 ft/min
f_{max}	0.005 in/rev		

21.5 EVALUATIVE QUESTIONS

1. In a problem with two constraints, how many solutions are possible? Describe the conditions under which the solutions would occur.

2. The data for the metal cutting problem has been modified and is in the Table 21.2. Calculate the values for the constants and variables.

3. The data for the metal cutting problem has been modified and is in the Table 21.3. Calculate the values for the constants and variables. However, for this problem consider the horsepower to be non-binding. Calculate the horsepower used since it is non-binding.

Table 21.2: Data for Problem 2

Input Parameters		Calculated Constants and Variables	
Q	1	K_{00}	————
R_0	0.04 \$/min	K_{01}	————
R_m	0.06 \$/min	K_{02}	————
C_t	2.0 \$/edge	K_{11}	200
t_h	1.5 min	K_{21}	0.358
t_{ch}	0.80 min	Z	————
D	1 inch	ω_{01}	————
L	6 inches	ω_{02}	————
m	0.80	ω_{11}	————
n	0.25	ω_{21}	————
C	$5.00 * 10^8$ min	Y	1.14
a	3.58	C_u	————
b	0.91	B	————
c	0.78	f	0.005 in/rev
Hp	10	V	290 ft/min
f_{max}	0.005		

Table 21.3: Data for Problem 3 (horsepower constraint removed)

Input Parameters		Calculated Constants and Variables	
Q	1	K_{00}	————
R_0	0.04 \$/min	K_{01}	————
R_m	0.06 \$/min	K_{02}	————
C_t	2.0 \$/min	K_{11}	200
t_h	1.5 min	K_{21}	0.358
t_{ch}	0.80 min	Z	————
D	1 inch	ω_{01}	————
L	6 inches	ω_{02}	————
m	0.80	ω_{11}	————
n	0.25	ω_{21}	————
C	$5.00 * 10^8$ min	Y	0.91
a	3.58	C_u	————
b	0.91	B	————
c	0.78	f	0.005 in/rev
Hp	NA	V	460 ft/min
f_{max}	0.005		

21.6   REFERENCES

[1] R. C. Creese and Pingfang Tsai, Generalized solution for constrained metal cutting economics problem, *1985 Annual International Industrial Conference Proceedings*, pp. 113–117, Institute of Industrial Engineers U.S. 141

[2] D. S. Ermer, Optimization of the constrained machining economics problem by geometric programming, *Journal of Engineering for Industry*, pp. 1067–1072, Transactions of the ASME, 1971. 141, 148

[3] T. Pingfang, *An Optimization Algorithm and Economic Analysis for a Constrained Machining Model*, p. 214, Ph.D. Dissertation, West Virginia University, Morgantown, WV.

C H A P T E R   22

The Open Cargo Shipping Box with Skids Case Study

22.1   INTRODUCTION

The open cargo shipping box was presented in Chapter 7 and is the classical problem in geometric programming. It had zero degrees of difficulty and was solved relatively easily. The open cargo shipping box problem is adjusted to add skid rails at $5/unit length, and if two rails are used, the additional cost would be $10/box length. This becomes a problem with one degree of difficulty and the solution is more difficult, and the purpose of this chapter is to introduce different methods for obtaining a solution.

The first method to be applied is the *constrained derivative approach*, that is obtaining four of the dual variables in terms of the fifth variable, taking the derivative of the function and setting it to zero and obtaining a fifth independent dual equation. The problem then becomes a problem with zero degrees of difficulty and the solution is obtained.

The second method applied is the *dimensional analysis approach*, which takes the primal-dual equations and separates them into the primal variables on one side of the equation and the dual variables and constants on the other side. Solving this for the exponents of the terms leads to the required additional independent equation. The second method does not require the taking of the derivatives of logarithmic functions which can be difficult. This method is similar to the *substitution approach* in developing additional equations. One advantage of the dimensional analysis method is that students are often familiar with this technique in evaluating units engineering problems.

The third method is the *condensation of terms approach*. This method combines two terms to reduce the degrees of difficulty by one. This method, however, does not guarantee an optimal solution and the selection of which terms to combine is important. The terms selected should have exponents that are near in value and should be selected such that the equations in the new dual do not result in dual variables becoming zero or that redundant equations result in the formation of the dual. The terms are typically combined geometrically with equal weights.

22.2   PRIMAL AND DUAL PROBLEM FORMULATION

The problem can be stated as: "400 cubic yards of gravel must be ferried across a shallow river. The box is an open box with skids used because of the low water level. The box has length L, width

W, and depth H. The sides of the box cost \$10/yd² and have a total area of $2(L + W)H$ and the bottom of the box cost \$20/yd² and has an area of LW. There are 2 skids at a cost of \$5/yd on the length of the box and the total skid length is $2L$. Each round trip of the box on the ferry will cost 10 cents per cubic yard of gravel shipped." The objective function becomes:

$$C = 40/LWH + 10LW + 20LH + 40HW + 10L. \tag{22.1}$$

The previous cost solution was \$100 without skids and the box length was 2 units (yds), so the new cost would be \$100 + \$20 = \$120 if the same dimensions are used. However, one must determine if the additional cost yields the same dimensions ($L = 2$ yds, $W = 1$ yd, $H = 1/2$ yd) and 400 trips were required.

The primal objective function can be written in general terms of:

$$C = C_{01}/(LWH) + C_{02}LW + C_{03}LH + C_{04}HW + C_{05}L, \tag{22.2}$$

where

$$C_{01} = 40$$
$$C_{02} = 10$$
$$C_{03} = 20$$
$$C_{04} = 40$$
$$C_{05} = 10.$$

The degree of difficulty is:

$$D = T - (N + 1) = 5 - (3 + 1) = 1, \tag{22.3}$$

where

$$T = \text{number of terms}$$
$$N = \text{number of variables.}$$

The dual formulation is:

Objective Function	$\omega_{01} + \omega_{02} + \omega_{03} + \omega_{04} + \omega_{05} = 1$	(22.4)
L terms	$-\omega_{01} + \omega_{02} + \omega_{03} \quad\quad +\omega_{05} = 0$	(22.5)
W terms	$-\omega_{01} + \omega_{02} \quad\quad + \omega_{04} \quad\quad = 0$	(22.6)
H terms	$-\omega_{01} \quad\quad +\omega_{03} + \omega_{04} \quad\quad = 0.$	(22.7)

Subtracting Equation (22.7) from Equation (22.6) one obtains that:

$$\omega_{02} = \omega_{03}. \tag{22.8}$$

Considering Equation (22.4) and Equation (22.5), one obtains

$$\omega_{04} = 1 - 2\omega_{01}. \tag{22.9}$$

Using Equation (22.6) with Equations (22.7) and (22.8), one obtains

$$\omega_{02} = 3\omega_{01} - 1, \tag{22.10}$$

and using Equation (22.4) with Equations (22.8) thru (22.10), one obtains

$$\omega_{05} = 2 - 5\omega_{01}. \tag{22.11}$$

The dual can be written as:

$$\text{Dual}\,(Y) = \{(C_{01}/\omega_{01})^{\omega_{01}} (C_{02}/\omega_{02})^{\omega_{02}} (C_{03}/\omega_{03})^{\omega_{03}}$$
$$(C_{04}/\omega_{04})^{\omega_{04}} (C_{05}/\omega_{05})^{\omega_{05}}\} \tag{22.12}$$

$$Y = \left\{(C_{01}/\omega_{01})^{\omega_{01}} (C_{02}/(3\omega_{01} - 1))^{(3\omega_{01}-1)} (C_{03}/(3\omega_{01} - 1))^{(3\omega_{01}-1)}\right.$$
$$\left.(C_{04}/(1 - 2\omega_{01}))^{(1-2\omega_{01})} (C_{05}/(2 - 5\omega_{01}))^{(2-5\omega_{01})}\right\}. \tag{22.13}$$

22.3 CONSTRAINED DERIVATIVE APPROACH

The constrained derivative approach takes the derivative of the log Y with respect to ω_{01}, sets it to zero, and solves for ω_{01}.

$$\text{Log}Y = \omega_{01} \log(C_{01}/\omega_{01}) + (3\omega_{01} - 1) \log(C_{02}/(3\omega_{01} - 1))$$
$$+ (3\omega_{01} - 1) \log(C_{03}/(3\omega_{01} - 1)) + (1 - 2\omega_{01})$$
$$\log(C_{04}/(1 - 2\omega_{01})) + (2 - 5\omega_{01}) \log(C_{05}/(2 - 5\omega_{01})) \tag{22.14}$$

$$\partial(\log Y)/\partial(\omega_{01}) = 0 = -1 + \log(C_{01}/\omega_{01}) + -3 + 3\log(C_{02}/(3\omega_{01} - 1))$$
$$- 3 + 3\log(C_{03}/(3\omega_{01} - 1)) + 2 - 2\log(C_{04}/(1 - 2\omega_{01}))$$
$$+ 5 - 5\log(C_{05}/(2 - 5\omega_{01})) \tag{22.15}$$

$$= \log(C_{01}/\omega_{01}) + 3\log(C_{02}/(3\omega_{01} - 1)) + 3\log(C_{03}/(3\omega_{01} - 1))$$
$$- 2\log(C_{04}/(1 - 2\omega_{01})) - 5\log(C_{05}/(2 - 5\omega_{01})). \tag{22.16}$$

The antilog is taken and the constants are put on one side and the dual variable equations on the other to obtain:

$$C_{01}C_{02}^3 C_{03}^3/(C_{04}^2 C_{05}^5) = Z$$
$$= \omega_{01}(3\omega_{01} - 1)^3 (3\omega_{01} - 1)^3/(1 - 2\omega_{01})^2 (2 - 5\omega_{01})^5 \tag{22.17}$$

$$Z = 40 \times 10^3 \times 20^3/(40^2 \times 10^5) = 2 \tag{22.18}$$

$$\omega_{01}(3\omega_{01} - 1)^3 (3\omega_{01} - 1)^3/(1 - 2\omega_{01})^2 (2 - 5\omega_{01})^5 = 2. \tag{22.19}$$

Note that the terms of Equation (22.19) must be positive, so that indicates that ω_{01} must be < 1, $> 1/3$, < 0.5, and < 0.4 which indicates it is between $1/3$ and 0.4. If one solves Equation (22.19) for ω_{01} (using search techniques), the value obtained was:

$$\omega_{01} = 0.3776519. \tag{22.20}$$

Using Equations (22.8) to (22.11) for the other variables in terms of ω_{01}, one obtains

$$\omega_{01} = 0.378 \tag{22.21}$$
$$\omega_{02} = 0.133 \tag{22.22}$$
$$\omega_{03} = 0.133 \tag{22.23}$$
$$\omega_{04} = 0.245 \tag{22.24}$$
$$\omega_{05} = 0.111. \tag{22.25}$$

Using the values of C and ω in the Equation (22.12) for the dual, one obtains

$$Y = (40/0.378)^{0.378}(10/0.133)^{0.133}(20/0.133)^{0.133}$$
$$(40/0.245)^{0.245}(10/0.111)^{0.111} = \$115.72. \tag{22.26}$$

This compares with the cost of \$100 for the initial problem in Chapter 7 or the estimated cost of \$120 when the skid cost is added to the non-skid box. Using the primal-dual relationships and the objective function, the primal variables can be found as:

$$L = \omega_{05}Y/C_{05} = (.111)(115.72)/10 = 1.284 \text{ yd (vs. 2 yd).} \tag{22.27}$$

Similarly,

$$H = \omega_{03}Y/(C_{03}L) = \omega_{03}Y/(20\omega_{05}Y/10) = (\omega_{03}/\omega_{05})(C_{05}/C_{03})$$
$$= (0.133/0.111)(10/20) = 0.599 \text{ yd (vs. 1.0 yd),} \tag{22.28}$$

and

$$W = \omega_{02}Y/(C_{02}L) = \omega_{02}Y/(10\omega_{05}Y/10) = (\omega_{02}/\omega_{05})(C_{05}/C_{02})$$
$$= (0.133/0.111)(10/10) = 1.198 \text{ yd (vs. 1.0 yd).} \tag{22.29}$$

Therefore, the primal becomes

$$C = C_{01}/(LWH) + C_{02}LW + C_{03}LH + C_{04}HW + C_{05}L \tag{22.30}$$
$$= C_{01}/(1.284 * 0.599 * 1.198) + C_{02}(1.284 * 1.198) + C_{03}(1.284 * 0.599)$$
$$+ C_{04}(0.599 * 1.198) + C_{05} * (1.284)$$
$$= 43.41 + 15.38 + 15.38 + 28.70 + 12.84$$
$$= 115.71 \text{ versus } 115.72. \tag{22.31}$$

The volume of the box is $1.284 \times 0.599 \times 1.198 = 0.921$ cubic yards vs. the 1.0 cubic yard volume in the original problem. The number of trips will increase from 400 in the original problem to 434.3 or 435 trips. The shipping cost increases from \$40 to \$43.41, but the box cost (side, end, and bottom) totals \$59.46 vs. \$60 and the primary increase is the cost of the skids, that is \$12.84.

22.4 DIMENSIONAL ANALYSIS APPROACH FOR ADDITIONAL EQUATION

The dimensional analysis approach starts with the primal and dual equations which are:

$$C = C_{01}/(LWH) + C_{02}LW + C_{03}LH + C_{04}HW + C_{05}L \tag{22.2}$$

and

$$\text{Dual}(Y) = \{(C_{01}/\omega_{01})^{\omega_{01}}(C_{02}/\omega_{02})^{\omega_{02}}(C_{03}/\omega_{03})^{\omega_{03}}$$
$$(C_{04}/\omega_{04})^{\omega_{04}}(C_{05}/\omega_{05})^{\omega_{05}}\} . \tag{22.12}$$

The primal dual relationships are:

$$L^{-1}W^{-1}H^{-1} = \omega_{01}Y/C_{01} \tag{22.32}$$
$$LW = \omega_{02}Y/C_{02} \tag{22.33}$$
$$LH = \omega_{03}Y/C_{03} \tag{22.34}$$
$$HW = \omega_{04}Y/C_{04} \tag{22.35}$$
$$L = \omega_{03}Y/C_{03}. \tag{22.36}$$

These relationships can be combined to give:

$$(L^{-1}W^{-1}H^{-1})^A(LW)^B(LH)^C(HW)^D(L)^E = 1 = (\omega_{01}Y/C_{01})^A(\omega_{02}Y/C_{02})^B$$
$$(\omega_{03}Y/C_{03})^C(\omega_{04}Y/C_{04})^D(\omega_{05}Y/C_{05})^E. \tag{22.37}$$

The equations from dimensional analysis are:

L terms	$-A + B + C \quad\quad + E = 0$	(22.38)
W terms	$-A + B \quad\quad + D \quad\quad = 0$	(22.39)
H terms	$-A \quad\quad + C + D \quad\quad = 0$	(22.40)
Y	$+A + B + C + D + E = 0.$	(22.41)

From Equations (22.39) and (22.40), one obtains that:

$$B = C. \tag{22.42}$$

Subtracting Equation (22.38) from Equation (22.41) one obtains:

$$D = -2A. \tag{22.43}$$

Subtracting Equation (22.38) from Equation (22.39) results in

$$D = C + E. \tag{22.44}$$

From Equation (22.38) one observes that:

$$C + E = A - B. \tag{22.45}$$

Thus,

$$D = -2A = A - B \tag{22.46}$$

so

$$B = 3A \tag{22.47}$$
$$C = 3A \tag{22.48}$$
$$D = -2A, \tag{22.49}$$

and via Equation (22.38)

$$E = -5A. \tag{22.50}$$

Thus, if $A = 1$, then $B = 3$, $C = 3$, $D = -2$, and $E = -5$.

Thus, from Equation (22.37)

$$1 = (\omega_{01} Y / C_{01})^1 (\omega_{02} Y / C_{02})^3 (\omega_{03} Y / C_{03})^3 (\omega_{04} Y / C_{04})^{-2} (\omega_{05} Y / C_{05})^{-5}. \tag{22.51}$$

This results in

$$\begin{aligned}(\omega_{01})(\omega_{02})^3(\omega_{03})^3(\omega_{01})^{-2}(\omega_{01})^{-5} &= (C_{01})(C_{02})^3(C_{03})^3(C_{04})^{-2}(C_{05})^{-5} \\ &= (40)(2)^3(10)^3(40)^{-1}(10)^{-5} \end{aligned} \tag{22.52}$$

or in terms of ω_{01}

$$(\omega_{01})(3\omega_{01} - 1)^3(3\omega_{01} - 1)^3(1 - 2\omega_{01})^{-2}(2 - 5\omega_{01})^5 = 2. \tag{22.53}$$

Equation (22.53) is the same as Equation (22.19) and the solution procedure from that point on would be the same as that used for the constrained derivative approach.

22.5   CONDENSATION OF TERMS APPROACH

The technique of condensation involves the combining of terms to reduce the degrees of difficulty to make the solution easier. D. J. Wilde presented this technique in his book *Globally Optimum Design* in 1978 and this example is presented in *Engineering Design—A Material and Processing Approach* by George Dieter (1991).

The primal problem was:

$$C = 40/LWH + 10LW + 20LH + 40HW + 10L \tag{22.1}$$

or as

$$C = C_{01}/(LWH) + C_{02}LW + C_{03}LH + C_{04}HW + C_{05}L. \tag{22.2}$$

The dual is

$$\text{Dual}\,(Y) = \{(C_{01}/\omega_{01})^{\omega_{01}}(C_{02}/\omega_{02})^{\omega_{02}}(C_{03}/\omega_{03})^{\omega_{03}}$$
$$(C_{04}/\omega_{04})^{\omega_{04}}(C_{05}/\omega_{05})^{\omega_{05}}\}, \tag{22.12}$$

and there is one degree of difficulty.

The condensation technique combines two terms to reduce the degrees of difficulty. The selection of the terms and the weighing of the terms is important. The terms selected should have exponents that are not that different and should be selected such that equations in the new dual do not result in dual variables resulting to be zero or that redundant equations result in the formulation of the dual. The weights for combining the two terms will be considered to be equal. The formation of the new term will be illustrated.

If the third and fifth terms are combined with equal weights (1/2), the result is

$$t_3 + t_5 \geq (C_{03}LH/(1/2))^{1/2}(C_{05}L/(1/2))^{1/2} = 2(C_{03}C_{05})^{1/2}LH^{1/2} = C_{06}LH^{1/2}, \tag{22.54}$$

where

$$C_{06} = 2(C_{03}C_{05})^{1/2}. \tag{22.55}$$

The new primal would be:

$$C = C_{01}/(LWH) + C_{02}LW + C_{04}HW + C_{06}LH^{1/2}. \tag{22.56}$$

The new dual would be

$$\text{Dual}\,(Y) = \{(C_{01}/\omega_{01})^{\omega_{01}}(C_{02}/\omega_{02})^{\omega_{02}}(C_{04}/\omega_{04})^{\omega_{04}}(C_{06}/\omega_{06})^{\omega_{06}}\}. \tag{22.57}$$

The new dual formulation is:

Objective Function	$\omega_{01} + \omega_{02} + \omega_{04} + \omega_{06} = 1$	(22.58)
L terms	$-\omega_{01} + \omega_{02} \qquad\quad + \omega_{06} = 0$	(22.59)
W terms	$-\omega_{01} + \omega_{02} + \omega_{04} \qquad = 0$	(22.60)
H terms	$-\omega_{01} \qquad + \omega_{04} + \omega_{06}/2 = 0.$	(22.61)

Equations (22.59) and (22.60) result in:

$$\omega_{06} = \omega_{04}. \tag{22.62}$$

Equation (22.61) and (22.62) result in:

$$\omega_{01} = \omega_{04} + \omega_{06}/2 = (3/2)\omega_{04}. \tag{22.63}$$

Using Equations (22.62) and (22.63) in Equation (22.59) results in:

$$\omega_{02} = \omega_{01} - \omega_{06} = \omega_{04}/2. \tag{22.64}$$

Using Equations (22.62), (22.63), and (22.64) with Equation (22.58) results in:

$$\omega_{04} = 0.25 \tag{22.65}$$
$$\omega_{06} = 0.25 \tag{22.66}$$
$$\omega_{01} = 0.375 \tag{22.67}$$
$$\omega_{02} = 0.125. \tag{22.68}$$

Using the values for the constants and the dual variables, the dual objective function becomes:

$$\text{Dual}\,(Y) = \left\{ (40/0.375)^{0.375}(10/0.125)^{0.125}(40/0.25)^{0.25}(2(20 \times 10)^{1/2}/0.25)^{0.25} \right\}$$
$$= (5.761)(1.729)(3.557)(3.261) = 115.5(\text{versus } 115.71). \tag{22.69}$$

The primal dual relationships yield

$$C_{01}/HWL = \omega_{01}Y \tag{22.70}$$
$$C_{02}LW = \omega_{02}Y \tag{22.71}$$
$$C_{04}HW = \omega_{04}Y \tag{22.72}$$
$$C_{06}LH^{1/2} = \omega_{06}Y. \tag{22.73}$$

Combining Equations (22.71) and (22.72) results in:

$$H/L = (C_{02}/C_{04})(\omega_{04}/\omega_{02}) = (10/40)(0.25/0.125) = 1/2. \tag{22.74}$$

Combining Equations (22.71) and (22.73) results in:

$$(W/H^{1/2}) = (C_{06}/C_{02})(\omega_{02}/\omega_{06})$$
$$= (2(20 \times 10)^{1/2}/(10) \times (0.125/0.25) = 2^{1/2} = 1.414. \tag{22.75}$$

Combining Equations (22.70) and (22.71) results in

$$L^2W^2H = (C_{01}/C_{02})(\omega_{02}/\omega_{01}) = (40/10)(0.125/0.375) = 4/3. \tag{22.76}$$

Using Equation (22.74) and (22.75) with Equation (22.76) one obtains:

$$(2H)^2(2H)H = 4/3, \qquad (22.77)$$

and this results in:

$$H = (1/6)^{0.25} = 0.6389 \text{ yd} \quad \text{(vs. 0.599 yd in constrained derivative solution)} \qquad (22.78)$$
$$L = 2H = 1.2779 \text{ yd} \quad \text{(vs. 1.284 yd in constrained derivative solution)} \qquad (22.79)$$
$$W = (2H)^{1/2} = 1.1304 \text{ yd} \quad \text{(vs. 1.198 yd in constrained derivative solution).} \qquad (22.80)$$

The primal objective function would be:

$$C = C_{01}/(LWH) + C_{02}LW + C_{04}HW + C_{06}LH^{1/2} \qquad (22.81)$$
$$= 40/(1.2779 \times 1.1304 \times 0.6389) + 10(1.2779 \times 1.1304)$$
$$+ 40(0.6389 \times 1.1304) + 20 \times 2^{1/2} \left(1.2779 \times 0.6389^{1/2} \right)$$
$$= 43.34 + 14.45 + 28.89 + 28.89$$
$$= 115.57. \qquad (22.82)$$

The values for the primal and dual are in good agreement with each other and in good agreement with the constrained derivative approach and the dimensional analysis approach. The primal variables for the condensed version are slightly different than the primal variables of the other solutions. This is because the objective function is slightly different, but the values of the various objective functions are quite close.

22.6 EVALUATIVE QUESTIONS

1. If the cost of the skid rails was $20 instead of $10, what is the effect on the total cost, the number of trips, and the box dimensions?

2. Use the dimensional analysis approach on the metal cutting problem with two constraints to derive Equation (21.31).

3. Use the condensation technique with second and fifth terms of the primal objective function and compare the solutions of the original problem and when the condensation technique was used on terms 3 and 5.

4. What are the box dimensions and total cost if the transportation cost C_{01} is increased to $80 (increase cost to 0.20/yd^3)?

22.7 REFERENCES

[1] D. J. Wilde, *Globally Optimum Design*, Wiley-Interscience, NY, pp. 88–90, 1978.

[2] G. E. Dieter, *Engineering Design: A Materials and Processing Approach*, pp. 223–225, 2nd Ed., McGraw-Hill, NY, 1991.

CHAPTER 23

Profit Maximization Considering Decreasing Cost Functions of Inventory Policy Case Study

23.1 INTRODUCTION

The classical inventory models consider the unit costs and unit prices to be fixed, but as the quantity increases the economies of scale permit decreases in costs and prices. In the model presented by Jung and Klein [1] in 2001, they considered cost per unit and price per unit to be power functions of demand. The variables in the profit maximization model considered are the order quantity and the product demand. The inventory model considered is the basic model with the assumptions that (1) replenishment is instantaneous; (2) no shortage is allowed; and (3) the order quantity is a batch.

23.2 MODEL FORMULATION

The variables and parameters used in the model are listed in Table 23.1. The variables are demand per unit time and order quantity. The parameter price is a function of the price scaling constant, the demand, and the price elasticity constant. The parameter cost is a function of the cost scaling constant, the demand, and the cost elasticity constant, also called the economy of scale factor.

The symbols used for the variables and parameters are presented in Table 23.1 The profit maximization model can be stated on a per unit time basis as:

$$\text{Maximize Profit} (\pi) = \text{Total Variable Revenue} (R) - [\text{Total Variable Cost (TVC)}$$
$$+ \text{ Total Set-up Cost (TSC)} + \text{Inventory Holding Cost (IHC)}].$$

$$(23.1)$$

The revenue per unit can be represented as:

$$P = aD^{-\alpha}, \tag{23.2}$$

Table 23.1: Variables and parameters for inventory model

Symbol Used	Description of Variable or Parameter
D	Demand per Unit Time (decision variable)
Q	Order Quantity (decision variable)
P	Price per Unit ($/Unit)
C	Cost per Unit ($/Unit)
A	Set-up Cost ($/Batch)
i	Inventory Holding Cost Rate (Percent/Unit Time)
a	Scaling Constant for Price (Initial Price)
b	Scaling Constant for Cost (Initial Cost)
α	Price Elasticity with respect to Demand
β	Cost Elasticity with respect to Demand (Degree of Economies of Scale)
π	Profit per unit time
R	Total Revenue
TVC	Total Variable Cost
TSC	Total Set-up Cost
IHC	Total Inventory Holding Cost

$$D = \text{Demand per Unit Time}$$

where a = Scaling Constant for Price

α = Price Elasticity with Respect to Demand.

The total revenue (R) is the product of the demand and the revenue per unit:

$$R = D * P = D * aD^{-\alpha} = aD^{1-\alpha}. \tag{23.3}$$

The cost per unit (C) can be represented as:

$$C = bD^{-\beta}, \tag{23.4}$$

where

D = Demand per Unit Time

b = Scaling Constant for Cost

β = Cost Elasticity with Respect to Demand.

The total variable cost (TVC) is the product of the demand and the cost per unit:

$$TVC = D * C = D * bD^{-\beta} = bD^{1-\beta}. \tag{23.5}$$

The total set-up cost (TSC) is the product of the set-up cost times the number of set-ups which can be expressed as:

$$TSC = A * D/Q = ADQ^{-1}. \tag{23.6}$$

The inventory holding cost (IHC) represents the product of the average inventory ($Q/2$), the unit cost ($bD^{-\beta}$), and the inventory holding cost rate (i) and results in:

$$IHC = Q/2 * bD^{-\beta} * i = (ib/2) * QD^{-\beta}. \tag{23.7}$$

Inserting Equations (23.3), (23.5), (23.6), and (23.7) into Equation (23.1), the result is the primal objective function:

$$\text{Max}\,(\pi) = aD^{1-\alpha} - \left(bD^{1-\beta} + ADQ^{-1} + (ib/2)QD^{-\beta}\right). \tag{23.8}$$

The problem can be written in general terms, such as

$$\text{Max}\,(\pi) = C_{oo}D^{1-\alpha} - C_{01}D^{1-\beta} - C_{02}DQ^{-1} - C_{03}QD^{-\beta}. \tag{23.9}$$

To solve the problem one minimizes the negative of the profit function, that is, the primal objective function becomes:

$$\text{Min}\,Y = -C_{oo}D^{1-\alpha} + C_{01}D^{1-\beta} + C_{02}DQ^{-1} + C_{03}QD^{-\beta}, \tag{23.10}$$

where $Y = -\pi$.

The degree of difficulty (D) is:

$$\text{Degrees of Difficulty} = T - (N + 1) = 4 - (2 + 1) = 1, \tag{23.11}$$

where

T = number of terms

N = number of variables

The signum functions would be:

$$\sigma_{01} = -1$$
$$\sigma_{02} = 1$$
$$\sigma_{03} = 1$$
$$\sigma_{04} = 1,$$

and $\sigma_{00} = -1$ as this is a maximization problem.

The dual objective function is:

$$D(Y) = -1\left[(C_{00}/\omega_{01})^{-\omega_{01}}(C_{01}/\omega_{02})^{\omega_{02}}(C_{02}/\omega_{03})^{\omega_{03}}(C_{03}/\omega_{04})^{\omega_{04}}\right]^{-1}. \tag{23.12}$$

The dual formulation would be:

Objective Function	$-\omega_{01} + \quad \omega_{02} + \omega_{03} + \quad \omega_{04} \quad = -1$	(23.13)
D Terms	$-(1-\alpha)\omega_{01} + (1-\beta)\omega_{02} + \omega_{03} - \beta\omega_{04} \quad = \quad 0$	(23.14)
Q Terms	$-\omega_{03} \quad + \omega_{04} \quad = \quad 0$	(23.15)

Since the degree of difficulty is 1, one additional equation is needed to solve the problem. First one must get the four dual variables in terms of one variable and then obtain an equation in terms of that variable. If one examines Equation (23.15) one observes that:

$$\omega_{03} = \omega_{04}. \tag{23.16}$$

Thus, one selects ω_{04} as the unknown variable and now must find ω_{01} and ω_{02} in terms of ω_{04}. If one multiplies Equation (23.13) by $(1-\alpha)$ and subtracts it from Equation (23.14) and using Equation (23.16), one obtains:

$$\omega_{02}[(1-\beta) - (1-\alpha)] + \omega_{04}[(1-\beta) - 2(1-\alpha)] = 1 - \alpha. \tag{23.17}$$

This results in:

$$\omega_{02} = [(1-\alpha)/(\alpha-\beta)] + [(1+\beta-2\alpha)/(\alpha-\beta)]\omega_{04}. \tag{23.18}$$

Now if one solves for ω_{01} via Equations (23.13) and (23.16) one obtains:

$$\omega_{01} = 1 + \omega_{02} + 2\omega_{04}. \tag{23.19}$$

Now using Equation (23.18) in Equation (23.19), one obtains the expression for ω_{01} in terms of ω_{04} which is:

$$\omega_{01} = [(1-\beta)/(\alpha-\beta)][1 + \omega_{04}]. \tag{23.20}$$

To obtain the additional equation, the approach of dimensional analysis will be used. The additional equations from the primal dual relationships are:

$$C_{00}D^{1-\alpha} = \omega_{01}Y \tag{23.21}$$

$$C_{01}D^{1-\beta} = \omega_{02}Y \tag{23.22}$$

$$C_{02}DQ^{-1} = \omega_{03}Y \tag{23.23}$$

$$C_{03}D^{-\beta}Q = \omega_{04}Y. \tag{23.24}$$

The dimensional analysis equation is:

$$(D^{1-\alpha})^{-A}(D^{1-\beta})^{B}(DQ^{-1})^{C}(D^{-\beta}Q)^{D} = 1 = (\omega_{01}Y/C_{00})^{-A}(\omega_{02}Y/C_{01})^{B}$$
$$(\omega_{03}Y/C_{02})^{C}(\omega_{04}Y/C_{03})^{D}. \tag{23.25}$$

Since this is a maximization problem, the sign on the revenue term exponent is negative and the three cost terms is positive. The equations would be:

$$D \text{ terms} \quad -A(1-\alpha) + B(1-\beta) + C - \beta D = 0 \tag{23.26}$$

$$Q \text{ terms} \quad -C + D = 0 \tag{23.27}$$

$$Y \text{ Dual} \quad -A \quad + B \quad + C + D = 0. \tag{23.28}$$

From Equation (23.27) one observes that

$$C = D. \tag{23.29}$$

If one multiplies Equation (23.28) by $(1-\alpha)$ and subtracts Equation (23.26) and using Equation (23.29) one obtains:

$$B = [(1+\beta-2\alpha)/(\alpha-\beta)]D. \tag{23.30}$$

Using Equation (23.28) one obtains that

$$A = B + C + D.$$

Using the values for B and C in terms of D the result is:

$$A = [(1-\beta)/(\alpha-\beta)]D. \tag{23.31}$$

If one selects the value of $D = 1$, the values are $A = [(1-\beta)/(\alpha-\beta)]$, $B = [(1+\beta-2\alpha)/(\alpha-\beta)]$, and $C = 1$. Now Equation (23.25) becomes:

$$(\omega_{01}Y/C_{00})^{-[(1-\beta)/(\alpha-\beta)]}(\omega_{02}Y/C_{01})^{[(1+\beta-2\alpha)/(\alpha-\beta)]}(\omega_{03}Y/C_{02})(\omega_{04}Y/C_{03}) = 1. \tag{23.32}$$

Separating the constants and dual variables, the relationship obtained was:

$$\omega_{01}^{-[(1-\beta)/(\alpha-\beta)]}\omega_{02}^{[(1+\beta-2\alpha)/(\alpha-\beta)]}\omega_{03}\omega_{04} = C_{00}^{-[(1-\beta)/(\alpha-\beta)]}$$
$$C_{01}^{[(1+\beta-2\alpha)/(\alpha-\beta)]}C_{02}C_{03} = K. \tag{23.33}$$

The right-hand side of the equation consists of only constants, so the product is also a constant. If one substitutes the dual variables in terms of ω_{04}, the expression becomes quite complex and would need to be solved with specific values of the parameters. The equation would be:

$$[((1-\beta)/(\alpha-\beta))(1+\omega_{04})]^{-[(1-\beta)/(\alpha-\beta)]}$$
$$[(1-\alpha)/(\alpha-\beta) + (1+\beta-2\alpha)/(\alpha-\beta)\omega_{04}]^{[(1+\beta-2\alpha)/(\alpha-\beta)]}\omega_{04}\omega_{04} = K. \tag{23.34}$$

23.3 INVENTORY EXAMPLE PROBLEM WITH SCALING CONSTANTS FOR PRICE AND COST

Elasticity

The input parameters for the example are in Table 23.2. Using the values of Table 23.2 in Equation (23.34), the result is:

$$[2.25(1 + \omega_{04})]^{-2.25}[1.25 + 0.25\omega_{04}]^{0.25}\omega_{04}^2 = 1.4058 \times 10^{-4}. \qquad (23.35)$$

Table 23.2: Parameters for illustrative problem

Symbol	Value	Description of Variable or Parameter
A	$10	Set-up Cost ($/Batch)
i	0.10	Inventory Holding Cost Rate (Decimal Percent/Unit Time)
a	200	Scaling Constant for Price (Initial Price)
b	20	Scaling Constant for Cost (Initial Cost)
α	0.5	Price Elasticity with respect to Demand
β	0.1	Cost Elasticity with respect to Demand (Degree of Economies of Scale)
C_{00}	200	$C_{00} = a$
C_{01}	20	$C_{01} = b$
C_{02}	10	$C_{02} = A$
C_{03}	1	$C_{03} = ib/2$

Solving 23.35 via search techniques for ω_{04} the result is:

$$\omega_{04} = 0.0296. \qquad (23.36)$$

Using this value for ω_{04}, the values of the remaining dual variables can be found from Equation (23.16), (23.18), and (23.20) as:

$$\omega_{03} = \omega_{04} = 0.0296 \qquad (23.37)$$
$$\omega_{02} = [(1 - \alpha)/(\alpha - \beta) + ((1 + \beta - 2\alpha)/(\alpha - \beta))\omega_{04}]$$
$$= [1.25 + 0.25 * (0.0296] = 1.2574 \qquad (23.38)$$
$$\omega_{01} = [((1 - \beta)/(\alpha - \beta))(1 + \omega_{04})] = 2.25 * (1 + 0.0296) = 2.3166. \qquad (23.39)$$

The value of the dual objective function is found from Equation (23.12) as:

$$D(Y) = -1\left[(C_{00}/\omega_{01})^{\omega_{01}}(C_{01}/\omega_{02})^{\omega_{02}}(C_{02}/\omega_{03})^{\omega_{03}}(C_{03}/\omega_{04})^{\omega_{04}}\right]^{-1} \qquad (23.12)$$

$$= -\left[(200/2.3166)^{-2.3166}(20/1.2574)^{1.2574}(10/0.0296)^{0.0296}(1/0.0296)^{0.0296}\right]^{-1}$$

$$= -[0.0013982]^{-1}$$

$$= -715.2. \qquad (23.40)$$

The minus sign indicates that it is a profit and not a cost. It is also interesting to note from the dual variables that the first cost term is the dominant cost term (95.50%) and that the second and third terms have equal value (2.25%). The difference between the revenue dual variable and the sum of the cost dual variables is 1.0, which is the total of the dual variables when a cost-only model is used.

The primal variables can be obtained from the dual variables using the primal-dual relationships. Rearranging the primal-dual relationship of Equation (18.20), the demand D can be evaluated as:

$$D = (\omega_{01}Y/C_{00})^{(1/(1-\alpha))} \qquad (23.41)$$

$$= (2.3166 * 715.18/200)^{(1/0.5)}$$

$$= 68.6. \qquad (23.42)$$

The value for the order quantity Q can be determined rearranging Equation (23.23) as:

$$Q = ((C_{02}D)/(\omega_{03}Y)) \qquad (23.43)$$

$$= ((10 * 68.6)/(0.0296 * 715.18))$$

$$= 32.4. \qquad (23.44)$$

The value of the primal objective function can now be found using Equation (23.10)

$$\text{Min } Y = -C_{oo}D^{1-\alpha} + C_{01}D^{1-\beta} + C_{02}DQ^{-1} + C_{03}QD^{-\beta} \qquad (23.10)$$

$$= -200(68.6)^{0.5} + 20(68.6)^{0.9} + 10 * 68.6/32.4 + 1 * 32.4(68.2)^{-0.1}$$

$$= -1656.5 + 898.9 + 21.2 + 21.2$$

$$= -715.2. \qquad (23.45)$$

The values of the primal and dual are in agreement. This problem was also solved by Yi Fang [2] via the transformed dual approach in which a constraint is added and the problem is solved via the constrained derivative approach, as the degree of difficulty is one. The results are identical, although the solution approaches are quite different, as the method presented obtains the additional equation via the method of dimensional analysis and solves a problem with zero degrees of difficulty.

23.4   TRANSFORMED DUAL APPROACH

The objective function of the primal is given by Equation (23.8) and is repeated here:

$$\text{Max}(\pi) = aD^{1-\alpha} - (bD^{1-\beta} + ADQ^{-1} + (ib/2)QD^{-\beta}). \tag{23.8}$$

Since the objective function was a signomial problem, Jung and Klein [1] used the transformed dual method developed by Duffin et al. [3] to solve the problem, and their solution is presented. The problem transforms the primal objective function into a constraint with positive terms. The problem was formulated as:

$$\text{Max } z, \tag{23.46}$$

subject to:

$$aD^{1-\alpha} - bD^{1-\beta} - ADQ^{-1} - (ib/2)QD^{-\beta} \geq z. \tag{23.47}$$

The transformed primal problem is:

$$\text{Min } z^{-1}, \tag{23.48}$$

subject to:

$$a^{-1}D^{\alpha-1}z + a^{-1}bD^{\alpha-\beta} + a^{-1}AD^{\alpha}Q^{-1} + a^{-1}(ib/2)QD^{\alpha-\beta-1} \leq 1. \tag{23.49}$$

The transformed primal function is a constrained posynomial function. The degree of difficulty remains the same as one variable and one terms was added, that is:

$$D = T - (N + 1) = 5 - (3 + 1) = 1. \tag{23.50}$$

From the coefficients and signs, the signum functions for the dual are:

$$\sigma_{01} = 1$$
$$\sigma_{11} = 1$$
$$\sigma_{12} = 1$$
$$\sigma_{13} = 1$$
$$\sigma_{14} = 1$$
$$\sigma_{1} = 1.$$

The dual problem formulation is:

Obj. Function ω_{01}	$= 1$	(23.51)
z terms $\quad -\omega_{01} \quad + \omega_{11}$	$= 0$	(23.52)
D terms $\quad + (\alpha - 1)\omega_{11} + (\alpha - \beta)\omega_{12} + \alpha\omega_{13} \quad +(\alpha - \beta - 1)\omega_{14} = 0$		(23.53)
Q terms $\quad\quad\quad\quad\quad\quad\quad\quad - \omega_{13} \quad\quad\quad\quad + \omega_{14} = 0.$		(23.54)

Using Equations (23.51)–(23.54) the values of the dual variables ω_{01} and ω_{11} can be found and ω_{12} can be found in terms of ω_{14}, and ω_{13} can be found in terms of ω_{14}. The results are:

$$\omega_{01} = 1 \tag{23.55}$$

$$\omega_{11} = 1 \tag{23.56}$$

$$\omega_{12} = [(1 - \alpha + (1 + \beta - 2\alpha)\omega_{14})/(\alpha - \beta)] \tag{23.57}$$

$$\omega_{13} = \omega_{14} \tag{23.58}$$

and

$$\omega_{10} = \omega_{11} + \omega_{12} + \omega_{13} + \omega_{14}$$

or

$$\omega_{10} = 1 + [(1 - \alpha - (2\alpha - \beta - 1)\omega_{14})/(\alpha - \beta)] + 2\omega_{14}$$
$$= ((1 - \beta)(1 + \omega_{14}))/(\alpha - \beta)). \tag{23.59}$$

The dual objective function can be stated in terms of ω_{14} as:

$$D(Y) = \left[(1/\omega_{01})^{\omega_{01}} \left(a^{-1}\omega_{10}/\omega_{11} \right)^{\omega_{11}} \left(a^{-1}b\omega_{10}/\omega_{12} \right)^{\omega_{12}} \right.$$
$$\left. \left(a^{-1}A\omega_{10}/\omega_{13} \right)^{\omega_{13}} \left((a^{-1}ib/2)\,\omega_{10}/\omega_{14} \right)^{\omega_{14}} \right]^{-1}, \tag{23.60}$$

which, upon substitution, becomes:

$$D(Y) = [1]^1 * \left[a^{-1} * ((1-\beta)(1+\omega_{14}))/(\alpha-\beta)) \right]^1$$
$$* \left[[(a^{-1}b) * ((1-\beta)(1+\omega_{14}))/(\alpha-\beta)] / \right.$$
$$\{1 - \alpha + (1 + \beta - 2\alpha)\omega_{14}/(\alpha-\beta)\}]^{\{(1-\alpha+(1+\beta-2\alpha)\omega_{14}/(\alpha-\beta)\}}$$
$$* \left[(a^{-1}A * ((1-\beta)(1+\omega_{14})/(\omega_{14}(\alpha-\beta))^{\omega_{14}} \right]$$
$$* \left[(a^{-1}(ib/2) * (1-\beta)(1+\omega_{14})/(\omega_{14}(\alpha-\beta))^{\omega_{14}}) \right]. \tag{23.61}$$

The logarithm form of Equation (23.61) is:

$$\text{Log}\,[D(Y)] = \log\left[(a^{-1}(1+\beta)/(\alpha-\beta)) * (1+\omega_{14}) \right]$$
$$+ \{(1 - \alpha + (1 + \beta - 2\alpha)\omega_{14}/(\alpha-\beta)\} \log\left[(a^{-1}b) \right.$$
$$* ((1-\beta)(1+\omega_{14}))/(\alpha-\beta))/\{1 - \alpha + (1 + \beta - 2\alpha)\omega_{14}/(\alpha-\beta)\}]$$
$$+ \omega_{14} \log\left(a^{-1}A * (1-\beta)(1+\omega_{14}) \right) / (\omega_{14}(\alpha-\beta))$$
$$+ \omega_{14} \log\left(a^{-1}(ib/2) * (1-\beta)(1+\omega_{14}) \right) / (\omega_{14}(\alpha-\beta)). \tag{23.62}$$

Taking the derivative of Equation (23.62) and setting it to zero and reducing the terms results in:

$$
\begin{aligned}
&\text{Log}\left\{\left[(a^{-1}b)(1-\beta)(1+\omega_{14})\right]/[1-\alpha+(1+\beta-2\alpha)\omega_{14}]\right\}^{(1+\beta-2\alpha)/(\alpha-\beta)} \\
&+ \text{Log}\left[a^{-1}A((1-\beta)/(\alpha-\beta))*((1+\omega_{14})/\omega_{14})\right] \\
&+ \text{Log}\left[a^{-1}(ib/2)((1-\beta)/(\alpha-\beta))*((1+\omega_{14})/\omega_{14})\right] = 0.
\end{aligned}
\tag{23.63}
$$

Taking the anti-log this becomes:

$$
\begin{aligned}
&\left\{\left[(a^{-1}b)(1-\beta)(1+\omega_{14})\right]/[1-\alpha+(1+\beta-2\alpha)\omega_{14}]\right\}^{(1+\beta-2\alpha)/(\alpha-\beta)} \\
&* \left[a^{-1}A((1-\beta)/(\alpha-\beta))*((1+\omega_{14})/\omega_{14})\right] \\
&* \left[a^{-1}(ib/2)((1-\beta)/(\alpha-\beta))*((1+\omega_{14})/\omega_{14})\right] = 1.
\end{aligned}
\tag{23.64}
$$

Using the values for the example problem of $\alpha = 0.5$, $\beta = 0.1$, $a = 200$, $b = 20$, $A = 10$, and $i = 0.10$ to solve for ω_{14}, Equation (23.64) becomes

$$
\begin{aligned}
&\left\{\left[(200^{-1}20)(1-0.1)(1+\omega_{14})\right]/[1-0.5+(1+0.1-2*0.5)*\omega_{14}]\right\}^{[(1+0.1-2*0.5)/(0.5-0.1)]} \\
&* \left[(200^{-1}*10*((1-0.1)/(0.5-0.1))*((1+\omega_{14})/\omega_{14}))\right] \\
&* \left[(200^{-1}*(0.1*20/2)*((1-0.1)/(0.5-0.1))*((1+\omega_{14})/\omega_{14}))\right] = 1
\end{aligned}
\tag{23.65}
$$

which reduces to

$$
\begin{aligned}
&[0.09*(1+\omega_{14})]/[0.5+0.1*\omega_{14}]^{0.25}*[0.1125*(1+\omega_{14})/(\omega_{14})] \\
&* [0.01125*(1+\omega_{14})(\omega_{14})] = 1.
\end{aligned}
\tag{23.66}
$$

Solving for ω_{14} by search techniques one obtains:

$$
\omega_{14} = 0.0296.
\tag{23.67}
$$

This is the same result as obtained previously in Equation (23.36) for ω_{04}, and thus the solutions will be the same as that of the dimensional analysis approach used. The constraint dual variables are the same as the objective dual variables as the constraint was equivalent to the earlier objective function. The remaining dual variables are:

$\omega_{12} = 1.2574$

    (same as ω_{02} earlier, but represents the same function in this problem)   (23.68)

$\omega_{13} = 0.0296$

    (same as ω_{03} earlier, but represents the same function in this problem)   (23.69)

$\omega_{11} = 1.0$ (23.70)

$\omega_{10} = 2.3167.$ (23.71)

The value of the dual objective function using Equation (23.60) is

$$
D(Y) = \left[(1/\omega_{01})^{\omega_{01}} \left(a^{-1}\omega_{10}/\omega_{11} \right)^{\omega_{11}} \left(a^{-1}b\omega_{10}/\omega_{12} \right)^{\omega_{12}} \right.
$$
$$
\left. \left(a^{-1}A\omega_{10}/\omega_{13} \right)^{\omega_{13}} \left(\left(a^{-1}ib/2 \right) \omega_{10}/\omega_{14} \right)^{\omega_{14}} \right]^{-1} \tag{23.60}
$$
$$
= \left[(1//1)^1 * \left(200^{-1} * 2.3174/1 \right)^1 * \left(200^{-1} * 20 * 2.3167/1.2574 \right)^{1.2574} \right.
$$
$$
* \left(200^{-1} * 10 * 2.3167/0.0296 \right)^{0.0296}
$$
$$
\left. * \left(200^{-1} * (0.1 * 20/2) * 2.3167/0.0296 \right)^{0.0296} \right]^{-1}
$$
$$
= \left[1.3978 * 10^{-3} \right]^{-1}
$$
$$
= 715.4. \tag{23.72}
$$

The dual solution is in agreement with the dual solution obtained using the dimensional analysis approach. The solution for the primal variables can be obtained from the primal-dual relationships. The equations used to solve for the dual variables were quite different, but gave the same value for ω_{14} and its equivalent of ω_{04} of 0.0286. The approach using dimensional analysis was somewhat easier as it did not require taking the derivative of the log of the dual and it had one less dual variable than the transformed dual approach.

23.5 EVALUATIVE QUESTIONS

1. Solve the example problem changing the price elasticity to determine the effects of D and Q and compare the results with the original example problem.

Table 23.3: Parameters for Problem 1

Symbol	Value	Description of Variable or Parameter
A	$10	Set-up Cost ($/Batch)
i	0.10	Inventory Holding Cost Rate (Decimal Percent/Unit Time)
a	200	Scaling Constant for Price (Initial Price)
b	20	Scaling Constant for Cost (Initial Cost)
α	0.3	Price Elasticity with respect to Demand
β	0.1	Cost Elasticity with respect to Demand

2. Solve the example problem changing the set-up cost to determine the effects on D and Q and compare the results.

3. Solve for the primal variables using the dual variables obtained in the transformed dual approach example problem.

Table 23.4: Parameters for Problem 2

Symbol	Value	Description of Variable or Parameter
A	$80	Set-up Cost ($/Batch)
i	0.10	Inventory Holding Cost Rate (Decimal Percent/Unit Time)
a	200	Scaling Constant for Price (Initial Price)
b	20	Scaling Constant for Cost (Initial Cost)
α	0.5	Price Elasticity with respect to Demand
β	0.1	Cost Elasticity with respect to Demand

4. Take the derivative of Equation (23.61) and show all the steps to obtain Equation (23.66) including the terms that are canceled.

5. Discuss the advantages and disadvantages of the methods of dimensional analysis vs. the constrained derivative approach.

23.6 REFERENCES

[1] H. Jung, C. M. Klein, Optimal inventory policies under decreasing cost functions via geometric programming, *European Journal of Operational Research*, 132, pp. 628–642, 2001. 163, 170

[2] Y. Fang, IENG 593 F, *Geometric Programming Research Paper*, 23 pages, 2010. 169

[3] R. J. Duffin, E. L. Peterson, and C. Zener, *Geometric Programming—Theory and Application*, Wiley, NY, 1976. 170

PART V

Summary, Future Directions, Theses and Dissertations on Geometric Programming

CHAPTER 24

Summary and Future Directions

24.1 SUMMARY

The object of this text is to generate interest in geometric programming among manufacturing engineers, design engineers, manufacturing technologists, cost engineers, project managers, accountants, industrial consultants and finance managers by illustrating the procedure for solving certain industrial and practical problems utilizing geometric programming. The various case studies were selected to illustrate a variety of applications as well as a set of different types of problems from diverse fields. Several additional problems were added, focusing on profit maximization and additional problems with degrees of difficulty. In addition, the methods of dimensional analysis and the constrained derivative approach have been presented in detail to show how to generate additional equations when the degrees of difficulty are positive. Table 24.1 is a summary of the case studies presented in this text, giving the type of problem, degrees of difficulty, and other details.

The material removal/metal removal economics example and some Cobb-Douglas production function problems had variable exponents in the general solution. The problems were worked in detail so general solutions could be obtained and also to show that the dual and primal solutions were identical. The problems were selected to illustrate a variety of types and also to show the use of the primal-dual relationships to determine the equations for the primal variables. It is by showing the various types of applications in detailed examples that others can follow the procedure and develop new applications.

24.2 FUTURE DIRECTIONS

One major advantage of geometric programming is that the dual variables indicate the cost ratios of the primal expressions which indicate which terms to focus upon for cost reduction. The complexity of the primal is greatly reduced by the dual which permits solutions to complex problems more easily. The author is hopeful that others will offer additional examples to illustrate new applications that can be included in future editions. New applications will attract practitioners to this fascinating area of geometric programming.

It is believed that the scope of geometric programming will expand with new applications and new software programs. Software has been developed to solve geometric programming problems and complex problems can be solved, but programs for the development of the design equations has not been a focus of geometric programming research. New software is being de-

Table 24.1: Summary of case study problems

Chapter	Case Study	% of Difficulty	# of Constraints	# of Variables	Variable Description	# of Solutions	Special Characteristics
4	The optimal box design	0	1	3	Height, width, length	1	
5	Trash can	0	1	2	Height, diameter	1	
6	Building design	0	1	3	Height, length, width	1	
7	Open cargo shipping box	0	1	3	Height, width, length	1	Classical problem
8	Metal casting cylindrical riser	0	1	2	Height, diameter	1	
9	Inventory model	0	0	1	Lot size	1	
10	Process furnace design	0	1	3	Temperature, length, height	1	Dominant equation, negative dual variable
11	Gas transmission pipe line	0	1	4	Length, diameter, flow length pressure ratio factor	1	Four variables
12	Material removal/metal cutting	0	1	2	Feed rate, cutting speed	1	Variable exponents
13	Construction building sector	0	1	2	Labor, capital	1	Cobb-Douglas production function
14	Production function profit	0	0	2	Labor, capital	1	Cobb-Douglas production function
15	Profit profit maximization	0	0	3	Product 1, product 2, product 3	1	Cobb-Douglas production function
16	Chemical plant product profitability	0	0	4(3)	Temperature, pressure % catalyst, % product	1	Fixed relationship between two variables
17	Journal bearing design	1	1	2	Journal radius, bearing half-length	1	Dimensional analysis approach, substitution approach
18	Multistory building	1	1	2	Number of floors, length	1	Dimensional analysis approach, number of floors rounded to integer
19	Metal casting hemispherical top riser	2	1	2	Height, diameter	1	Dimensional analysis approach, substitution approach
20	Liquefied petroleum (LPG) cylinder	1	1	2	Height, diameter	2	Multiple solutions, dimensional anaylsis approach, substitution approach
21	Material removal/metal cutting economics-2 contraints	1	2	3	Feed rate, cutting speed	3	Multiple solutions, derivative approach
22	Open cargo shipping box with skids	1	0	3	Height, width, length	1	Derivative approach, dimensional analysis approach, condensation of terms approach
23	Profit maximizations with decreasing cost functions	1	0	2	Lot size, demand	1	Dimensional analysis approach,
		1	1	3	Lot size, demand	1	Transformed dual approach

veloped for solving *Generalized Geometric Programming Problems and Unconstrained Generalized Geometric Programming Problems*, which is tentatively scheduled to be released in the near future.

24.3   DEVELOPMENT OF NEW DESIGN RELATIONSHIPS

There are many different types of problems that can be solved by geometric programming and one of the significant advantages of the method is that it is possible in many applications to develop general design relationships. The general design relationships can save considerable time and effort in instances where the constants are changed. The potential to develop design relationships for the more complex problems could greatly improve product design and quality while reducing development time and product cost. Much more research is needed in the development of software to assist in obtaining design equations for the primal variables. The new developments in software will hopefully make the development of design relationships for the primal variables easier and faster.

CHAPTER 25

Theses and Dissertations on Geometric Programming

25.1 INTRODUCTION

A search was made to find the theses and dissertations on geometric programming using Proquest Dissertations Publishing, formerly known as Dissertation Abstracts Online. If a title appeared in both the M.S. and Ph.D. lists, it was listed only in the M.S. list. Four lists were developed and they are in the order starting with the earliest and ending with the latest. The four lists compiled were:

1. Table Master Theses with Geometric Programming in the Title

2. Master Theses with Geometric Programming in the Abstract, but not in the Title

3. Dissertations with Geometric Programming in the Title

4. Dissertations with Geometric Programming in the Abstract, but not in the Title.

25.2 LISTS OF M.S. THESES AND PH.D. DISSERTATIONS

There were ten master theses in the first list, but only one was after 2000. Although there were only five master theses in the second list, four of them were after 2000. This indicates that geometric programming is being recognized more recently as a tool for solving complex problems rather than as a new innovative topic.

The third and fourth lists include 72 dissertations with geometric programming in the title and 53 additional dissertations with geometric programming not listed in the title, but only listed in the abstract. However, since 2000, the number of dissertations with geometric programming listed only in the abstract was 27 ($\approx$ 50%) vs. only 10 ($\approx$ 14%) with geometric programming in the title. The earlier dissertations focused on improvements to the development of geometric programming theory and techniques for solving problems. The later dissertations have focused more on the applications of geometric programming to complex problems for solution.

Table 25.1: Master theses with geometric programming in thesis title

	Author	Title	School	Year
1	Shaw, Hemendrakumar R.	Optimal Heat-Exchanger Design by Geometric Programming	Univ. of Massachusetts–Lowell	1971
2	Shah, Shivji Virji	Optimization of Finned Tube Condenser by Geometric Programming	Univ. of Massachusetts–Lowell	1973
3	Yeh, Lucia Lung-Chia	Geometric Programming Solution of a Problem Involving Sampling from Overlapping Frames	University of Wyoming	1974
4	Yang, Edward V	Sensitivity of Optimum Heat Exchanger Design by Geometric Programming	Univ. of Massachusetts–Lowell	1976
5	Meter, James W.	A Geometric Programming Solution to a Strata Allocation Problem	University of Wyoming	1976
6	Carver, Leonard James	Application of Geometric Programming to PID Controller Tuning with State Constraints	The University of British Columbia	1976
7	Freedman, Paul	The Cost Minimization of Machining Operations Using Geometric Programming	The University of British Columbia	1983
8	Douglas-Jarvis, Patricia	A Geometric Programming Algorithm for Augmented-Zero Degree of Difficulty Problems	Lehigh University	1984
9	Liu, Gulli	The Hoelder Inequality and its Applications in Banach Algebra and Geometric Programming	The University of Regina	1992
10	Cheung, Wing Tai	Geometric Programming and Signal Flow Graph Assisted Design of Interconnect and Analog Circuits	University of Hong Kong	2008

Table 25.2: Master theses with geometric programming in the abstract but not in the title

	Author	Title	School	Year
1	Helba, Michael John	Development and Optimization of a Nonlinear Multi-bumper Design Model for Spacecraft Protective Structures Using Simulated Annealing	The University of Alabama at Huntsville	1994
2	Kokatnur, Ameet	Design and Development of a Target-Costing Model for Machining	West Virginia University	2004
3	Gupta, Deepak Prakash	Energy-Sensitive Machining Parameter Optimization Model	West Virginia University	2005
4	Rirouzabadi, Sina	Joint Optimal Placement and Power Allocation of Nodes of a Wireless Network	University of Maryland	2007
5	Wang, Chenyuan	Performance Evaluation and Enhancement for AF Two-Way Receiving in the Presence of Channel Estimation Error	University of Victoria	2012

Table 25.3: Dissertations with geometric programming in dissertation title (*Continues.*)

	Author	Title	School	Year
1	Passy, Ury	Generalization of Geometric Programming; Partial Control of Linear Inventory Systems	Stanford University	1966
2	Avriel, Mordecai	Topics in Optimization: Block Research: Applied and Stochastic Geometric Programming	Stanford University	1966
3	Joseph, George Ecker	Geometric Programming; Duality in Quadratic Programming and LP Approximation	The University of Michigan	1968
4	Kochenberger, Gary Austin	Geometric Programming; Extensions to Deal with Degrees of Difficulty and Loose Constraints	University of Colorado at Boulder	1969
5	Pascual, Luis, De La Cruz	Constrained Maximization of Posynomials and Vector-Valued Criteria in Geometric Programming	Northwestern University	1969
6	Staats, Gleen Edwin	Computational Aspects of Geometric Programming with Degrees of Difficulty	The University of Texas at Austin	1970
7	Teske, Clarence Eugene	Optimum Structural Design Using Geometric Programming	Texas Tech University	1970
8	Dinkel, John Joesph	A Duality Theory for Dynamic Programming Problems *Via* Geometric Programming	Northwestern University	1971
9	Oleson, Gary K.	Computational Aspects of Geometric Programming with Degrees of Difficulty	Iowa State University	1971
10	Unklesbay, Kenneth Boyd	The Extension of Geometric Programming to Optimal Mechanical Design	University of Missouri–Columbia	1971
11	McNamara, John Richard	The Optimal Design of Water Quality Management Systems: An Application of Multistage Geometric Programming	Rensselaer Polytechnic University	1971
12	Jefferson, Thomas Richard	Geometric Programming with Application to Transportation Planning	Northwestern University	1972
13	Beck, Paul Alan	A Modified Convex Simplex Algorithm for Geometric Programming with Subsidiary Problems	Rensselaer Polytechnic University	1972
14	McCarl, Bruce Alan	A Computational Study of Polynomial Geometric Programming	The Pennsylvania State University	1973
15	Smeers, Yves Marie	Geometric Programming with Applications to Management Science	Carnegie-Mellon University	1973
16	Vandrey, Joan Betty Wettern	Reversed Generalized Geometric Programming	Northwestern University	1973
17	Planchart, Alejo	A Geometric Programming Approach to the Problem of Finding Efficient Algorithms for Constrained Location Problems with Mixed Norms	Northwestern University	1973

Table 25.3: *(Continued.)* Dissertations with geometric programming in dissertation title *(Continues.)*

	Author	Title	School	Year
18	Mekaru, Mark Masanobu	N-Dimensional Transportation Problems: Algorithm for Linear Problems and Application of Geometric Programming for Non-linear Problems	Arizona State University	1973
19	Ben-Tal, Aharon	Contributions to Geometric Programming and Generalized Convexity	Northwestern University	1973
20	Weibking, Rolf D.	Deterministic and Stochastic Geometric Programming Models for Optimal Engineering Design in Electrical Power	Rensselaer Polytechnic University	1973
21	Emery, Sidney William, Jr.	Transoceanic Dry-Bulk Systems Optimization by Geometric Programming	Stanford University	1974
22	Salinas-Pacheco, Juan Jose	Minimum Cost Design of Concrete Beams Using Geometric Programming	University of Calgary	1974
23	Plecnik, Joseph Matthew	Optimization of Engineering Systems Using Extended Geometric Programming	The Ohio State University	1974
24	Zoracki, Michael Jon	A Primal Method for Geometric Programming	Rensselaer Polytechnic University	1974
25	Hall, Michael Anthony	A Geometric Programming Approach to Highway Network Equilibria	Northwestern University	1974
26	McMasters, John Hamilton	The Optimization of Low-Speed Flying Devices by Geometric Programming	Purdue University	1975
27	Kise, Peter Ethan	Systems of Optimization via Geometric Programming	University of Pennsylvania	1976
28	Gribik, Paul Robert	Semi-infinite Programming Equivalents and Solution Techniques for Optimal Experimental Design and Geometric Programming Problems with an Application to Environmental Protection	Carnegie-Mellon University	1976
29	Wu, Chin Chang	Design and Modeling of Solar Sea Power Plants by Geometric Programming	Carnegie-Mellon University	1976
30	Ramamurthy, Subramanian	Structural Optimization Using Geometric Programming	Cornell University	1977
31	Grange, Franklin Earnest, II	The Solution and Analysis of the Geometric Mean Portfolio Problem with Methods of Geometric Programming	Colorado School of Mines	1977
32	Maurer, Ruth Allene Lamb	A Geometric Programming Approach to the Preliminary Design of Wastewater Treatment Plants	Colorado School of Mines	1978
35	Hough, Clarence Lee, Jr	Optimization of the Second-Order Logarithmic Machining Economics Problem by Extended Geometric Programming	Texas A & M University	1978
36	Yale, Wilson Winant	The Design of Solid Waste Systems: An Application of Geometric Programming to Problems in Municipal Solid Waste Management	Lehigh University	1978
37	Rosenberg, Eric	Globally Convergent Algorithms for Convex Programming with Applications to Geometric Programming	Stanford University	1979

Table 25.3: *(Continued.)* Dissertations with geometric programming in dissertation title *(Continues.)*

	Author	Title	School	Year
38	Brar, Guri Singh	Geometric Programming for Engineering Design Optimization	The Ohio State University	1981
39	Mayer, Robert Hall, Jr	Cost Estimates from Stochastic Geometric Programs	University of Delaware	1982
40	Freedman, Paul	The Cost Minimization of Machining Operations Using Geometric Programming	The University of British Columbia	1983
41	Kyparisis, Jerzy	Sensitivity and Stability for Nonlinear and Geometric Programming; Theory and Applications	The George Washington University	1983
42	Ravenbakht, Camelia	Geometric Programming; An Efficient Computer Algorithm for Traffic Assignment (Network Equilibrium)	North Carolina State University	1984
43	Rajasekera, Jayantha Ranjith	Perturbational Techniques for the Solution of Posynomial, Quadratic, and L(P)-Approximation Programs (Geometric Programming, Nonlinear)	North Carolina State University	1984
44	Douglas-Jarvis, Patricia	A Geometric Programming Algorithm for Augmented-Zero Degree of Difficulty Problems	Lehigh University	1984
45	Djanali, Supeno	Geometric Programming and Decomposition Techniques in Optimal Control	University of Wisconsin	1984
46	Rajgopal, Jayant	A Duality Theory for Geometric Programming Based on Generalized Linear Programming	The University of Iowa	1985
47	Burns, Scott Allen	Structural Optimization Using Geometric Programming and the Integrated Formulation	University of Illinois at Urbana-Champaign	1985
48	Arreola Contreras, Jose Jesus	Semi-Discrete Geometric Programming	University of Pittsburgh	1986
49	Muhammed, Abdelfatah Adelmotti	Information Theory and Queuing Theory via Generalized Geometric Programming	North Carolina State University	1987
50	Choi, Jae Chul	Generalized Benders Decomposition for Geometric Programming with Several Discrete Variables	The University of Iowa	1987
51	Chibani, Lahbib	Integrated Optimal Design of Structures Subjected to Alternate Loads Using Geometric Programming	University of Illinois at Urbana-Champaign	1987
52	Kirk, John B.	A Geometric Programming Based Algorithm for Preprocessing Nonlinear Signomial Optimization Programs	Colorado School of Mines	1988
53	Wessels, Gysbert Johannes	A Geometric Programming Algorithm for Solving a Class of Nonlinear Signomial Optimization Problems	Colorado School of Mines	1989
54	Knowles, James Alesander	A Geometric Programming Approach to the Solution of Multistage Countercurrent Heat Exchanger Systems	Colorado School of Mines	1990

Table 25.3: *(Continued.)* Dissertations with geometric programming in dissertation title

	Author	Title	School	Year
55	Martino, Thomas John	A Tandem Joint Inspection and Queuing Model Employing Posynomial Geometric Programming	University of Rhode Island	1991
56	Katz, Joseph Harold	An Algorithm for Solving a Class of Nonlinear, Unconstrained Multi-Variable Signomial Optimization Problems Using Geometric Programming	Colorado School of Mines	1991
57	Dull, Owne S.	A Geometric Programming Based Method for Determining the Required Constraints in a Class of Chemical Blending Problems	Colorado School of Mines	1991
58	Cheng, Zhao-Yang	A Least Squared Approach to Interior Point Methods with an Application to Geometric Programming	Rensselaer Polytechnic University	1993
59	Jackson, Jack Allen, Jr.	A Mathematical Experiment in Dual Geometric Programming	Colorado School of Mines	1994
60	Yang, Hsu-Hao	Investigations of Path-Following Algorithms for Signomial Geometric Programming Problems	The University of Iowa	1994
61	Mader, Douglas Paul	An Improved Method for Sensitivity Analysis in Geometric Programming	Colorado School of Mines	1994
62	Ince, Erdem	A Parallel Balanced Posynomial Unconstrained Geometric Programming Algorithm	Colorado School of Mines	1995
63	Hershenson, Maria del Mar	CMOS Analog Circuit Design via Geometric Programming	Stanford University	2000
64	Chiang, Mung	Solving Nonlinear Problems in Communication Systems Using Geometric Programming and Dualities	Stanford University	2003
65	Colleran, David M.	Optimization of Phase-Locked Loop Circuits via Geometric Programming	Stanford University	2006
66	Hsiung, Kan-Lin	Geometric Programming under Uncertainty with Engineering Applications	Stanford University	2008
67	Cheung, Wing Tai	Geometric Programming and Signal Flow Graph Assisted Design of Interconnect and Analog Circuits	University of Hong Kong	2008
68	Joshi, Siddharth	Large-Scale Geometric Programming for Devices and Circuits	Stanford University	2008
69	Patil, Dinesh	Design of Robust Energy-Efficient Digital Circuits Using Geometric Programming	Stanford University	2008
70	Wang, Ya-Ping	Duality Theory for Composite Geometric Programming	University of Pittsburgh	2012
71	Xue, Siangzhong	Electronic System Optimization Design via GP-Based Surrogate Modeling	North Carolina State University	2012
72	Hoburg, Warren Woodrow	Aircraft Design Optimization as a Geometric Program	University of California, Berkeley	2013

Table 25.4: Dissertations with geometric programming in abstract but not in the title (*Continues.*)

	Author	Title	School	Year
1	Memon, Altaf Ahmed	Environmental Noise Management: An Optimization Approach	University of Pittsburgh	1980
2	Junna, Mohan Reddy	Irrigation System Improvement by Simulation and Optimization	Colorado State University	1980
3	Silberman, Gabriel Mauricio	The Design and Evaluation of an Active Memory Unit	State University of New York at Buffalo	1980
4	Ohanomah, Matthew Ochukoh	Computational Algorithms for Multi-component Phase Equilibria and Distillation	The University of British Columbia	1982
5	Martin Gracia, Angel	Contribution to the Mathematical Theory of Transport Planning and its Applications	Universidad Politecnica de Madrid	1982
6	Morningstar, Kay Gladys	Self-Teaching Enrichment Modules on Modern Applications of Mathematics for Talented Secondary Students and Evaluation of Effective Utilization Through an Original Prediction Mode	University of Delaware	1983
7	Duggar, Cynthia Miller	A Study of the Relationships Among Computer Programming Ability, Computer Program Content, Computer Programming Style and Mathematical Achievement in a College Level "Basic" Programming Course	Georgia State University	1983
8	Dziuban, Stephen T.	Ellipsoid Algorithm Variants in Nonlinear Programming	Rensselaer Polytechnic Institute	1983
9	Chou, Jaw Huoy	Contributions to Non-differential Mathematical Programming	North Carolina State University	1984
10	Tang, Chi-Chung	Mathematical Models and Optimization Techniques for use in Analysis and Design of Wastewater Treatment Systems	University of Illinois at Urbana-Champaign	1984
11	Murphy, Robert Craig	A Dual-Dual Algorithm for the Solution of a Class of Nonlinear Resource Allocation Models	Colorado School of Mines	1985
12	Tsai, Pingfang	An Optimization Algorithm and Economic Analysis for a Constrained Machining Model (Geometric Programming, Cost Optimization of Metal Cutting)	West Virginia University	1986
13	Plenert, Gerhard Johannes	A Solution to Smith's General Case Bottleneck Problem Allowing for an Unlimited Number of Scheduled Jobs	Colorado School of Mines	1987
14	Li, Xing-Si	Entropy and Optimization	The University of Liverpool	1987
15	Mrema, Godwill D.E.C.	Computational of Chemical and Phase Equilibria	Universitetet i Trondheim	1987
16	Thrme, James John	An Algorithm for Solving a Class of Nonlinear, Unconstrained, Signomial Economic Models Using the Greening Technique	Colorado School of Mines	1988
17	Bailey, Steven Scott	An Algorithm for the Solution of a Class of Economic Models for the Design of Pre-stressed Concrete Bridges	Colorado School of Mines	1989

Table 25.4: *(Continued.)* Dissertations with geometric programming in abstract but not in the title

	Author	Title	School	Year
18	No, Hoon	Accelerated Convergent Methods for a Class of Nonlinear Programming Problems	The University of Iowa	1990
19	Rama, Dasaratha Venkatara-man	Planning Models for Communication Satellites Offering Customer Premises Services	The University of Iowa	1990
20	Viriththamulla, Gamage Indrajith	Mathematical Programming Models and Heuristics for Standard Modular Design Problem	The University of Airzona	1991
21	Zhu, Jishan	On the Path-Following Methods for Linearly Constrained Convex Programming Problems	The University of Iowa	1992
22	Narang, Ramesh Vashoomal	Issues and Methodologies in Automating Process Planning from a Three-Dimensional Part Representation to Recommendation of Machining Parameters	The University of Iowa	1992
23	Maranas, Costas D.	Global Optimization in Computational Chemistry and Process Systems Engineering	Princeton University	1995
24	Zao, John Kar-kin	Finite-Precision Representation and Data Abstraction for Three-Dimensional Euclidean Transformations	Harvard University	1995
25	Xu, Li Na	Optimization Methods for Computing Empirically Constrained External Probability Distributions	The University of Iowa	1996
26	Lou, Jinan	Integrated Logical and Physical Optimization for Deep Submicron Circuits	University of Southern California	1999
27	Jung, Hoon	Optimal Inventory Policies for an Economic Order Quantity Model Under Various Cost Functions	University of Missouri - Columbia	2001
28	Jacobs, Etienne Theodorus	Power Dissipation and Timing in CMOS Circuits	Technische Universiteit Eindhoven	2001
29	Cheng, Hao	Theory and Algorithms for Cubic L(1) Splines	North Carolina State University	2002
30	Qin, Zhanhai	Topological Circuit Reduction: Theory and Applications	University of California, San Diego	2003
31	Xu, Yang	Affordable Analog and Radio Frequency Integrated Circuits Design and Optimization	Carnegie Mellon University	2004
32	O'Neil, Daniel C.	Jointly Optimal Network Performance: A Cross-Layer Approach	Stanford University	2004

Author's Biography

ROBERT C. CREESE

Dr. Robert C. Creese was Professor of Industrial and Management Systems Engineering at West Virginia University, USA and had taught courses on Engineering Economy, Advanced Engineering Economics, Cost and Estimating for Manufacturing, Manufacturing Processes and Advanced Manufacturing Processes. He has previously taught at The Pennsylvania State University (9 years), Grove City College (4 years), Aalborg University in Denmark (3 sabbaticals) and at West Virginia University for 35 years. He worked at US Steel for two years as an Industrial Engineer before starting his teaching career.

Dr. Creese is a Fellow of the Association for the Advancement of Cost Engineering, International (AACEI), received the Charles V. Keane Service Award and Brian D. Dunfield Educational Service Award presented by AACE, and was treasurer of the Northern West Virginia Section of AACE for more than 20 years. He is a Life Member of AACE International, ASEE (American Society for Engineering Education) and ASM (American Society for Materials). He also is a member of ICEAA (International Cost Estimating & Analysis Association) , AIST (Association for Iron & Steel Technology), AWS (American Welding Society), and AFS(American Foundry Society).

He obtained his B.S. Degree in Industrial Engineering from The Pennsylvania State University, his M.S. Degree in Industrial Engineering from the University of California at Berkeley, and his Ph.D. Degree in Metallurgy from The Pennsylvania State University.

Dr. Robert Creese has authored the book *Introduction to Manufacturing Processes and Materials* (Marcel Dekker-1999) and co-authored two books *Estimating and Costing for the Metal Manufacturing Industries*(Marcel Dekker-1992) with Dr. M. Adithan, Professor Emeritus of VIT University Vellore, India, and Dr. B.S. Pabla of the Technical Teachers' Training Institute, Chandigarh, India, and *Strategic Cost Analysis for Project Managers and Engineers* (New Age International Publishers-2010) with Dr. M. Adithan, VIT University, Vellore, India. He has authored/co-authored more than 100 technical papers.

Index

Printed in the United States
by Baker & Taylor Publisher Services